THE ADVENTURE OF ELEMENTS ON PLANET EARTH

VENTHAN NALATHAMBY

PARTRIDGE
A Penguin Random House Company

To order additional copies of this book, contact
Toll Free 800 101 2657 (Singapore)
Toll Free 1 800 81 7340 (Malaysia)
orders.singapore@partridgepublishing.com
www.partridgepublishing.com/singapore

To My Father
Nalathamby Muthan
And
For the Love of My Mother
Lechumey Linggem

PREFACE

The emergence of life on Earth has always fascinated Man. The awe of Space will also continue to inspire Man in his quest to find the origins of all cosmic existence.

Humankind had a definite beginning in the course of evolution on Earth but little does he know of his destiny. Undoubtedly now, Man is in the fore front of scientific knowledge and the next century will challenge him to conquer the shores of the Solar System and beyond.

Cosmologists and astrophysicists are able to unravel the secrets of the Universe and are even able to give vivid details of the Big Bang that happened billions of years ago. We are in the frontier of many new discoveries and the future to come will offer us fresh insight into the marvels of nature. In his relentless pursuit Man will soon unlock the mystery behind the emergence of life on Earth. Perception of life will profoundly change when new discoveries and knowledge throw light on our origin.

While scientific knowledge maybe within easy reach of the masses, scientific thoughts as such may not have taken strong hold in their perception of life. Prevalence of superstitious and illogical practises is not uncommon in many parts of the world even today! It is important to demystify non-scientific thoughts by way of presenting basic facts as they are. It is also necessarily essential to understand the basic sciences to appreciate Man's own existence in the natural scheme of things.

I believe that life on Earth is a result of interactions of chemicals which evolved into organic biological molecules. This book explains the basic chemistry of elements and atoms and how the very behaviour of elements naturally allowed the synthesis of biological compounds that inspired the emergence of life forms on Earth.

In this journey, we discover that all living things on Earth share the same common beginning and only one common molecular platform successfully emerged and all biodiversity on Earth springs from this one singular molecular platform.

The elements and molecules had a spectacular beginning from stardust, spewed out from explosions of stars billions of years ago, and the journey of the stardust, which was successful in making life forms on Earth, has to continue. Though the elements would have had different adventures in some other parts of the Universe, we owe it to ourselves to venture beyond the shores of Earth, the Solar System and the Milky Way Galaxy, with this successful molecular platform—DNA.

I sincerely hope that the readers of this book will find this work thought provoking, and be able to appreciate the beauty of molecular play.

I come from a family of ten siblings and I would like to thank all of them for their support especially my elder brothers, Muthusamy and Kumariah, who ignited my interest in the sciences. The environment at home, numerous discourses and discussions with my friends, teachers and fellow science mates have immensely helped me in conceptualising this book. Though many friends and relatives encouraged this work, I may not be able to mention all of them, but I would like to say thank you specifically to Yeoh Siok Kee for his support and encouragement. Special appreciation is due to G Sivapatham for availing himself for discussions on specific areas on biochemistry and to Mohd Redzuan for reading the first draft.

The illustrations in the book are the works of three undergraduates of LimKokWing University of Creative Technology, Cyberjaya, Malaysia. Ms. Mehreen Lamba, Ms. Danel Yessaliveva and Ms. Emau Saleem worked meticulously on all the illustrations and graphics in this book. Mehreen who comes from Patiala, a city in India known for its arts and culture, is pursuing her Bachelor of Design (Hons) in Professional Design (Visual Communication). Danel hails from Almaty, the commercial and cultural centre of Kazakhstan. Emau is from Male, the capital city of Maldives. Danel and Emau are working towards a Bachelor of Arts (Design) degree. Special thanks are due, to Mr. Timothy Vongsuthep, Programme Leader, Faculty of Design Innovation, for his efforts in organizing the undergraduates, and appreciation is also due to Ms Thilageswary Rajoo for her coordination work with LimKokWing University. I would also like to acknowledge Professor Emeritus Tan Sri Dato' Sri Dr Lim Kok Wing's celebrated passion for the arts and creativity, which led to the setting up LimKokWing University not only in Malaysia but also in many locations in the world. I sincerely believe the undergraduates of this prestigious institution will lead the next generation of creative thinkers.

Last but not the least, I wish to record my thanks to my wife Jacquilene for her support during the last two years and to my sons, Mannavan and Enaiavan for their incessant push for me to complete this book.

As I sign off this Preface, it is heartening to note that the rover Curiosity has made a successful touch down on Mars. We hope Curiosity will beam back crucial data on the possible existence of exobiological molecules in Mars. The rover's adventure on the surface of Mars will surely enhance our knowledge base and our understanding of our own earthly origin.

In any measure, it is a celebration for Humanity whenever there is a landing on any planet.

T Venthan Nalathamby
6 August 2012.
MALAYSIA

Note: The word Man is used in reference to Humankind.

CONTENTS

PART ONE
A JOURNEY FROM STARDUST
TO SELF DUPLICATING MOLECULES

PART TWO
THE SUCCESSFUL TRIAL THAT MADE LIFE POSSIBLE AND THE WAY BEYOND BIRTHPLACE EARTH

PART ONE

A JOURNEY FROM STARDUST TO SELF DUPLICATING MOLECULES

CHAPTER 1

The Common Beginning

The place: The Malay Archipelago.

I wondered why things were as they were while I was growing up as a young boy in the nineteen sixties. I grew up in a fishing village along a meandering river. My house was not too far away from the river bank leading towards the river mouth adjoining the Straits of Malacca. Days were care free as a young boy, and there was nothing else to do apart from roaming around the natural setting of mangrove swamp and coconut trees.

The river was a fascinating thing that affected everybody's life in this village town. Many depended on it for their livelihood. I had to cross the river by boat or open-top ferry to go to school. There were no bridges then. Leisure and playtime then were entirely out-door which were about looking for fighting fish and catching wild spiders. Nothing was like merely looking at the river and activities that went on there. I frequently walked along the river bank, and man-made bunds, looked at the muddy waters and was often greeted to the sight of mudskippers slithering back and forth from the river to the muddy banks.

The muddy river banks close to the river mouth formed a natural sanctuary for the mudskippers. The river water was greenish during high tides and brownish at low tides. The incessant switch of low and high tides created an ecosystem at the estuary, for the mudskippers and other creatures such as mud crabs and horse shoe crabs. They looked like creatures out of the world then.

Mangrove vegetation that swamped along the river mouth presented itself as an excellent playground for the mudskippers. It was an interesting past time for me to watch the tides come in to swell the river. The mud skippers disappeared into the waters along the river banks during high tides and showed up yet again at the next cycle of low tide. It was a joy seeing them playing and fighting with each other along the muddy river banks. Having had nothing better to do during those formative years, I developed a fascination for this creature and often caught them using strings made from plant material found in the mangrove swamps. I would play with the mudskippers for a while and then release them into the river as these creatures would dry up once they lose contact with water. They get dehydrated quickly and die if they are not released sooner into the waters. The mudskippers lived both on land and water but have to remain wet all the time.

As I grew older, I had this burning desire to find out the natural beginning of things both living and non-living within the world I lived. This passion became a relentless pursuit in my life's journey, and it remains so after all these years, even as I approach my fifty fifth birthday. The observable living things were startling enough. I was first introduced to the elements and atoms when I was 16 years of age. It took me quite a long while to understand and appreciate the nature of matter and existence of living things. It was, then, a mystery, as everything else. At that age, we take things for granted and accepted things as they are without question. As I got older, the intensity of this burning desire to find out what makes all these things tick grew even more and I had to find out why things exist as they are.

The world as we see through sensory eyes presents us a daily show of flourishing flora and fauna. They are all *full of sound and fury*. Earth is full of life and is full of drama performed by living things. Each day is like seeing real time drama being performed routinely and repeatedly by the biodiversity on Earth. We, as human beings witness the same drama, shown again and again over thousands of years. We are the only known audience of this drama. However, the show has been going on for millions of years even before Man's entry into the stage.

The storyline, in this drama, however, remains the same, except that the characters in the play continuously change with different features but using the same script. The drama is shown repeatedly over the entire course of time on Earth. But, there were no dull moments in this show. The show that we see now only represents a scene from, a long drawn drama, with no finishing script at sight. This outstanding spectacle on Earth is staged against a panoramic backdrop of a changing sky and vast starry expanse of the Cosmos.

The original stage was set on Earth more than 3 billion years ago. The show and dance was rehearsed over 3 billion years, and we still continue to see the rehearsals. Many characters come to perform their roles. Many left for good, and a few came back again, but many exited the stage permanently, never to return. But, the theme of the show goes on and on, adjusting itself to the changing environment at large. However, the grand finale of this largely impromptu drama is yet to unfold.

I like travelling especially when there is awe about it. In an adventurous journey, the aim is to arrive somewhere exciting, far away from home and perhaps halt for a while in some habitable place before continuing the journey again. The enjoyment is in the journey itself especially to unfamiliar destinations. To me, the essence of travel is to get out of the comfort zone, and if it to unchartered territory, then, it is even more exciting.

Sometimes, when we revisit a familiar place after a long lapse of time, we remember the past and trace back the history. I recently made a trip back to where I was born; to the fishing village, built along the river mouth of Selangor River. The casual visit back to my home town took me back many years into the past. I made it a point to visit the mangrove swamp. The mudskippers still visibly roamed the river beds. A host of other memories rushed into me, and for some reason, I remembered my class teacher Mr. Khoo Teng Yuen and his recital of poems in the 1970s. One such poem I vividly remembered was William Wordsworth's Daffodils. It is a pleasure to recall it time and again. Poems and songs are like time capsules, rekindles, memories.

William Wordsworth's Daffodils

I wandered lonely as a cloud
That floats on high o'er vales and hills,
When all at once I saw a crowd,
A host, of golden daffodils;
Beside the lake, beneath the trees,
Fluttering and dancing in the breeze.

Continuous as the stars that shine
And twinkle on the milky way,
They stretched in never-ending line
Along the margin of a bay:
Ten thousand saw I at a glance,
Tossing their heads in sprightly dance.

The waves beside them danced; but they
Out-did the sparkling waves in glee:
A poet could not but be gay,
In such a jocund company:
I gazed—and gazed—but little thought
What wealth the show to me had brought:

For oft, when on my couch I lie
In vacant or in pensive mood,
They flash upon that inward eye
Which is the bliss of solitude;
And then my heart with pleasure fills,
And dances with the daffodils.

Recollecting this poem while driving back from
Kuala Selangor, I did drift into thinking of the daffodils
which were likened to the stars that twinkle on the

Milky Way in Wordsworth's poem. I wonder what connection they all have in common, the mudskipper, the daffodils and the Milky Way.

Wordsworth's description of daffodils linking to the stars of Milky Way galaxy, I thought, was farfetched then, but I soon realised how closely they were related. As I know it now, they all share the same common beginning, both living and non-living.

But, where and how did it begin?

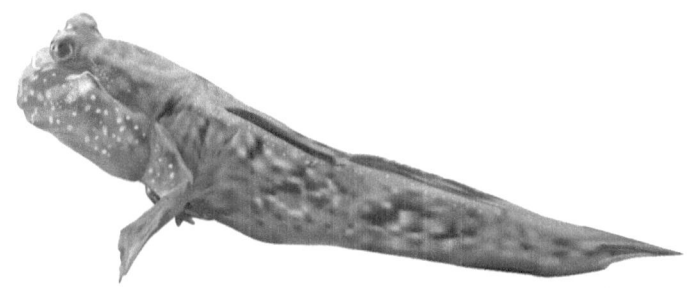

Illustration of Mudskipper

Footnote: Mudskipper is lungfish. It is able to live both in water and on land. Mudskipper belongs to Gobiidae family and sub family Oxudercinae. They are tropical and can be found in the equatorial regions and commonly in mangrove swamps. In water, they swim in side movement and on land they walk on pectoral fins or skip or jump.

Straits of Malacca, is between Peninsula Malaysia and Sumatra, (part of Indonesia). The fishing village is Pasir Penambang, and the classroom mentioned was in Sultan Abdul Aziz Secondary School, Kuala Selangor, Selangor, Malaysia.

CHAPTER 2

Start of a Long Journey

We live in this marvellous place call Earth-a round ball made of materials of sorts. The Earth, our planet revolves around a ball of fire and this ball of fire that we see in the day time is our Sun. There are eight planets that revolve around the Sun and Earth is the third nearest planet to it. This planet Earth is a home to life and living things, as we know it.

The Sun and the planets revolving around it, is referred to as the Solar System.

The Solar System is quietly tucked away in some obscure corner of a spiral galaxy we call Milky Way, amongst countless other galaxies. The relative position of Earth within the Solar System made it conducive for elements and molecules, to act out an astounding adventure.

The space beyond Earth is the rest of the Universe, which is estimated to contain a hundred billion galaxies. Milky Way is but one modest galaxy. A galaxy is a cluster of gases, cosmic dust and billions of stars harmoniously and violently co-existing. It is an enormous humongous structure of physics composed of debris and fragments which were built of cosmic material after a massive tremendous explosion. Each of these galaxies is estimated to have a spray of a hundred billion burning stars. It is just beyond ordinary minds to grasp. Not only the numbers are exceedingly large to deal with, the distance between these galaxies are also not earthly to comprehend.

To imagine distance, sometimes it is easier to think in terms of time taken to travel from one point to another. In many cultures, you would have heard of distance explained in terms of time. From where I came from, villagers, then, in the early 1960s would say that it would take quantifiable numbers of rolled cigarettes to reach a particular destination. A distance of one stick of cigarette is defined as the time taken to finish smoking one roll of cigarette. These villagers were obviously more familiar with cigarettes than distances measured in miles or yards. The distance between their home in the village and the local market was said to be 2 cigarettes then.

Astrophysicists, however, use time taken for light to travel to comprehend astronomical distances. Light, as in sunlight, travels at a speed of 186,000 miles or

300,000 kilo meters per second. In nature, speed of light is the fastest any particle can travel, and we have yet to see any undisputed evidence of particles moving beyond the speed of light. In astronomy, distance in space is better understood in units of light years. One light year, is the distance travelled by light for a period of one year. If a star is 4 light years away from Earth, then it simply means that if you travel at the speed of light, it will take 4 years to reach that star. Light travelling in a year would cover a distance of 10 trillion kilo meters or about 6 trillion miles.

The Milky Way Galaxy, our home galaxy, is a spiral disc like structure, has an approximate diameter of 100,000 light years and disc 'thickness' of around 1000 light years.

Our Universe, the Milky Way and Earth, did not exist in perpetuity. Earth has a history and a beginning, which can be traced back to some 4.6 billion years. Earth began as a fiery spherically ball which condensed from cosmic star dust gleaned from exploding stars. The Milky Way Galaxy and our local Solar System, owe its birth to the massive explosive event in cosmic time—space. The massive phenomenon is referred to as the Big Bang.

The Big Bang is the earliest record of an event signifying the beginning of the Universe. This singular cosmic event is estimated to have happened some 13.7 billion years ago.

As the name goes, Big Bang was an incident in astrophysical calendar that began with an explosion which created this Universe, galaxies, stars, black holes, quasar, planetary systems, time, space, asteroids, meteor, comets, matter, anti matter and black matter.

Big Bang represents an enormous, nuclear explosion of a primordial massive dense material. All postulations and theories point towards this beginning of enormous massive explosion and expansion. Billions of nuclear explosions spewed out the entire dense mass in random shapes and sizes sending asunder bits and pieces of it at speed, maybe even beyond that of light. The colossal masses resulting from this nuclear holocaust formed clusters as it cooled down under the laws of gravity and formed galaxies. It took about a billion years before galaxies formed, and they are still in the stage of violent formation resulting from the massive explosion. The Universe is still expanding as a result of the energy released from the mega nuclear blast off.

The Milky Way is still in its formative years. Along the margin of this Galaxy, our planetary system, Solar System formed some 5 billion years ago along with billions of other nuclear fired stars. The Milky Way is also home to some 400 billion stars. Within this nuclear inspired neighbourhood, Earth formed within the Solar System, located at one obscure corner in the outer spiralling arm of the revolving disc shaped Milky Way. Our Solar System is not even in the significant position in the Milky Way and is in no way near the centre which is some 30,000 light years away. Some obscure corner indeed is the address of our Solar System in our Galaxy.

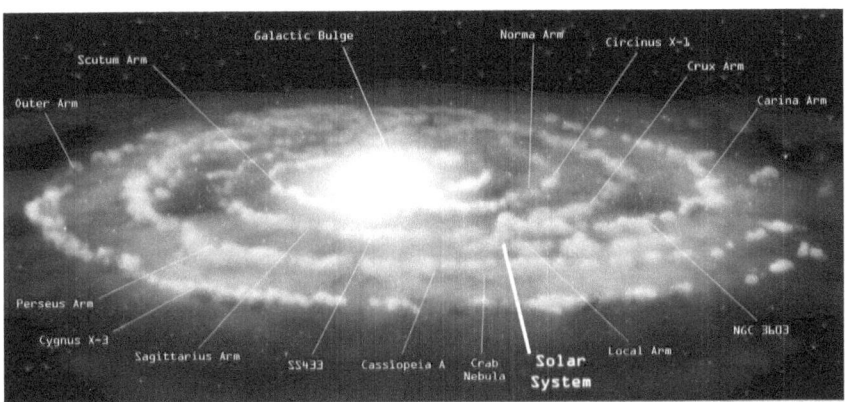

Milky Way and the relative position of the Solar System

The Sun, the centre of the Solar System is seen as a big ball of fire, reminding us the remnants of a mega off shoot of the Big Bang. The massive materials that spew out of the Big Bang are still cooling down. Our Sun, a nuclear power house, is about 330,000 times denser than Earth and more than a million times bigger than Earth, is a continuous and incessant furnace of nuclear fusion of Hydrogen in thermonuclear free play. The temperature there is about 10,000 degree F (water boils at 313 degree F). The Sun is 150 million kilo meter from Earth and light takes only about 8 minutes to reach Earth. The next nearest star from Earth, Proxima Centauri, is 4.2 light years away from

Earth. That is, if we can travel at the speed of light, it will take us 4 years 3 months to reach the nearest star.

The Sun, 8 light minutes away from Earth, is incredibly near in astronomical terms. The Sun, the source of energy, the omnipotent primal object in our neighbourhood space made the crucial difference in making many things possible on Earth.

The Earth took a long time to evolve itself to the current stage of look and feel as we see it today, weathering the harsh course of planet formation, journeying and sweeping through the vast spans of space. It survived the violent bombardments of debris, meteors and asteroids. During those formative years, the Earth went through extremely aggressive atmospheric and climatic changes. Earth itself is an unstable ball of solid and liquid molten rock. The various pieces of cooled Earth crust float on molten lava, some knocking against each other, some moving apart both on land and underneath the ocean floor. The violent eruptions of volcanoes are evidence of its aggressive beginning. The activities and the continuing murmur of Earth's core and tremble are humble reminder to us, now and then, through earth quakes, volcanic eruptions and tsunamis of its extreme beginning. The origin and birth of Earth is all telling through its rumples.

The Earth has evolved continuously and what we see today is the Earth that has survived the rough and tough cosmic world over the last 4.6 billion years.

The dust from exploding stars shaped the planets in the Solar System. The star dust provided the matters for the formation of Earth and other planets. The elements and substances found on Earth had a spectacular history which can be traced back to violent explosion of stars. As you would expect, Early Earth would have been terribly a different place from what it is today.

The Earth, today, can be deemed as a near perfect habitat for human living. But it does not mean that Earth was prepared for human dwelling. Human existence is quite recent in Earth's history. Earth formation had no specific design, and the course of Earth's path in arriving at today's environment was totally a random occurrence, a resultant of many interacting parameters, including geological and chemical interplay. The trail was totally unchartered. The Earth had a beginning with no timeline or milestones.

Time is a human invention, a tool to record the duration of passage of space on Earth and has no relevance to Universe space-time continuum. We need to understand the passage of space and use time as a relative index for our human comprehension.

A year, a unit of time period fixed by Man, is the time taken for Earth to complete one elliptical orbit around the Sun. So the time of 1 year is an Earth measured time with relevance to Man's existence. If Man did not exist, the concept of a year would not have occurred. Let us attempt a comprehensible framework to understand the passage of space and time.

We understand a passage of 100 years, the time taken for Earth to go around the Sun 100 times. Assuming the start of Big Bang till now is put into a timeframe of 100 years and if Big Bang is the beginning of year one, then the galaxies would have formed around year 6 and our local Sun would have formed 60 years after the Big Bang. The planetary system and our Earth would have begun to take shape around 70 years after the Bang.

Simple living single cell organism would have appeared only after 80 years and vertebrates would have surfaced after 99 years and Mammals would have been seen creeping into the Earth's ecosystem some 73 days before reaching the 100 year mark and *Homo sapiens* only coming in around the late evening before clinching the 100 year finishing line.

It undoubtedly took a long time for things to happen. No living form on Earth existed 80 % of the time, since Big Bang. Man did not exist for 99.9 % of the time since the beginning of time. Man is incidental in this overall scheme of the Big Bang.

It would be fascinating to know how the first life forms emerged during the course of such long time period on early Earth. The prototype of man only emerged some 5 million years ago and modern humans, *Homo sapiens* surfaced only about 200,000 years ago, and this present man grew to appreciate the natural wonders of Earth and the Universe. *Homo sapiens* is a culmination of many struggles of ordinary organisms that appeared on Earth before Man. *Homo sapiens* humble roots can be traced to every living form, including those in the shape of daffodils and mudskippers.

However, what made living things possible in the first place . . . and that too on Earth.

CHAPTER 3

It is Chemistry 101

Chemistry, Chemistry, Chemistry— What a Mystery

The daffodils and the mudskippers look so different and their means of survival poles apart, but today we know that both these organisms share the same biochemical platform that was passed on from very ancient common ancestors. The common baseline is not only applicable to these two organisms but to the rest of life forms on Earth, and this can be traced to chemistry and to the chemistry of Early Earth.

I have this passion for absolute knowledge, which I define, as knowledge that does not change with measured earthly time. Absolute knowledge is that knowledge, which can withstand the rigor of scientific inquiry before it is accepted as universal fact.

This work is my attempt to throw some light on how fundamental chemicals could have been instrumental for living things to emerge on Earth.

The journey into understanding the emergence of flora such as daffodils and fauna such as mudskippers begins with a peek at the nature of basic building components of these manifestations in biodiversity. The fundamental constituents of these matters are elements and molecules.

Emergence of living things had the origin to the attributes of elements and molecules. We need to comprehend the brick and mortar of elements and molecules in order to appreciate the emergence of living things on Earth. The path to understanding the existence of living things has to start with the chemistry of elements at its basic level.

Elements are pure substances that exhibit a particular behaviour consisting of only one type of atoms. Atom is a basic singular smallest unit that exhibits the behaviour of the element.

Elements are substances like Hydrogen, Oxygen, Carbon, Chlorine, Sodium, Calcium, Sulphur, Iron, Gold etc. The basic unit that would exhibit the characteristic of oxygen is the oxygen atom. Every element is unique, and its characteristic is uniform in nature. At the micro level, all elements are composed of atoms, and it is the atom that will be able to tell us the story of their behaviour and why they react the way they do, to produce different variety of substances.

Atoms are made up of electrons, protons and neutrons.

The behaviour of atoms in the chemical sense is attributed to the electrons in the atom. The very nature of electrons and its characteristic interaction with other electrons gave rise to molecules of different forms and shapes. Understanding the formation of organic molecules in living systems would throw light on how life would have sprung up on Earth. And, understanding atoms and electrons is the way to understanding the origin of life itself.

The journey into the world of atoms and electrons is as exciting as the journey probing the emergence of life on Earth. We will not go into serious chemistry of elements, but a basic understanding of atoms and element would be a compelling prelude to getting a glimpse into plausible explanation as to how these atoms would have come together to play out the greatest adventure that resulted in evolving life forms that also led to a species labelled as *Homo sapiens*—the modern thinking Human.

The chemistry of life begins at the atomic level. Understanding atoms and molecules is painless. Maybe dry but not necessarily an uninteresting one.

In simple terms, an atom is made up of positive, negative and neutral particles in a fuzzy enclosed minute space existing as an entity of its own. The negative particle is called electron, the positive particle is named proton. Neutrons are particles which are neutral. There are equal numbers of electrons and protons in an atom. Protons and neutrons give mass to an atom and occupy a nuclear core. The electrons in the atoms reside in fuzzy orbits surrounding the core. The electrons in the atom are almost weightless, and the entire mass of the atom comes from the nucleus comprising of proton and neutrons.

Atom = electron (negative charge) + proton (positive charge) + neutron (neutral charge)

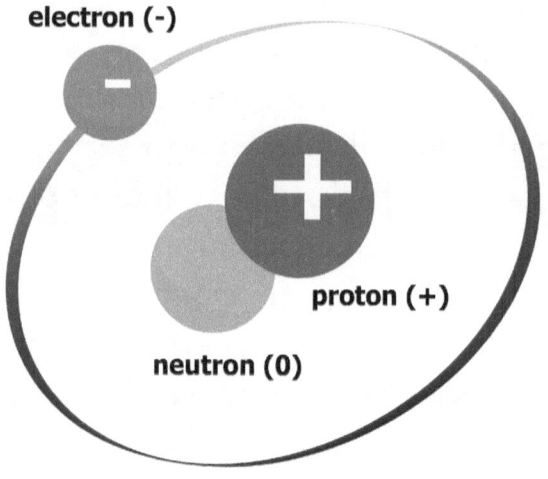

The simplest of an atom starts with one electron and one proton. Two or more atoms bind together to form what is called a molecule.

Atoms of different elements differ in the number electrons, and protons it contains. That's all the difference to it, but through this difference in the number of electron and proton, they exhibit considerably varying degree of properties. An addition of a single electron in an atom will make a world of difference to its behaviour.

The simplest of atoms with one electron in orbit and one proton in the core is Hydrogen. This atom occurs abundantly everywhere in the Universe, and this is the fundamental starting line for all other elements. This omnipresent element is the source of fuel for the stars including our Sun. Hydrogen is also the basic building block for the rest of the elements in the Universe. The fusion of two hydrogen atoms, give rise to another element, Helium.

The players in the creation of atoms are electrons, protons and neutrons and the configuration of these particles gave rise to different elements.

The origin of all elements can be traced to the stars. The cauldron of fire in the exploding stars provided the melting pot for electrons, protons and neutrons to build a systematic progression of atoms by simply adding on more electrons, protons and neutrons on to the simplest atom by way of fusion of existing atoms. The fusion of two Hydrogen atoms resulted in a Helium atom, which has two electrons, two protons and two neutrons. Progressively, new elements were created through fusion. It was a methodical exercise like an arithmetic progression giving rise to new elements in perfect sequence.

Let us theoretically do this exercise of methodically building up the elements from basic Hydrogen atom configuration. If we add another electron and proton to the Helium atom, we get Lithium, a metal that closely follows the behaviour of Sodium. If we add 3 more electrons to Hydrogen with a corresponding increase in protons and neutrons, we get Beryllium. Adding one more electron will result in 5 electrons, giving us Boron and yet one more additional electron to Boron will give us Carbon.

The chemical characteristic of an element is determined by the electrons and its spatial configuration in the atom. The electron arrangement and number of electrons in the atom are the deciding factors in determining the behaviour of the specific atom. The space occupied by electrons in an atom is called an orbit. The electron configuration in the orbits of an atom dictates the course of chemical reactions. The electrons 'arrange themselves' in a highly methodical manner in the orbits of the atom. The methodology of electron arrangement is the same throughout the Universe. It is specific and constant. And this is absolute.

The orbits that house the electrons have different energy levels. The electrons reside within these energy level orbits. The electrons arrange themselves in a packing order guided by energy bands, adhering to a strict rule of occupation.

The filling of electrons in the atomic orbital space is subject to definite rules, which are universal. There are different layers of orbits or shells. The number of orbits is dependent on the number of electrons that need to be housed. Inner orbits or shells are closer to the nuclear core and the outer shells are furthest away from the core. The outermost shell sits at the boundary end of the atom. Electrons are arranged from inner shells and proceed to fill outer shells as the number of the electron increases.

The first shell can accommodate only 2 electrons and the second shell can accommodate only 8 electrons, the third shell 8 and fourth and fifth 18 electrons each. The electron packing order is strictly followed. When the inner orbits are filled up, the additional electrons move to the next outer orbit. It is like filling the restaurant with a specific number of people fixed per table and following the sequence methodically. The table below shows the housing of electrons in the respective atoms.

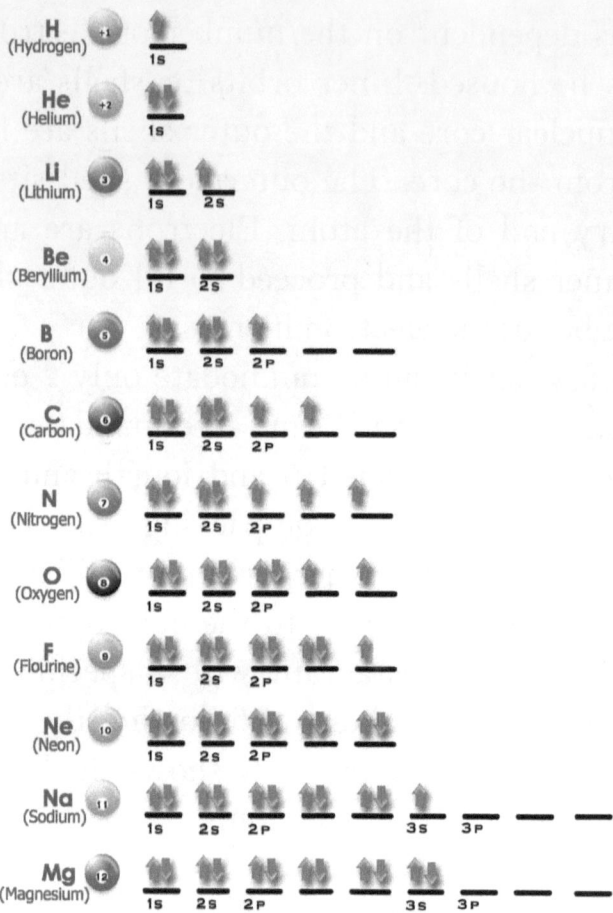

1s—first orbit closest to the atomic core

2s, 2p—second level orbits

3s, 3p—third level orbits

Electrons are the prime movers in determining the way the atoms behave. It is the world of electrons that bought us the array of chemicals with differing behaviour displaying a full spectrum in its adventure.

It is fundamentally relevant to look at the awesome character of these electrons that essentially propelled the atoms to associate amongst its comrades, to give us, the well organised diversity of chemicals. The peculiarity and the potential of this particle are amazing. Foremost, electron can be described as a wave or as a particle, as it can manifest itself either as a particle or as a wave under different circumstances to an outside observer like ourselves. The electron has dual face and chemists and physicists have learnt to live with it.

In an atom, each and every electron is different in terms of energy levels and the space it occupies. However, they are compelled to pair. When two electrons pair and occupy the same space or atomic shell, they distinguish themselves. They differentiate each other by spinning in an opposite direction. This universal rule is Pauli Exclusion Principle which simply says that no two electrons can occupy the same state simultaneously.

The coupling and pairing of electrons bring stability within the occupied space and to the atom as a whole. One other rule says that it is difficult to find out the electron's position and momentum at the same material time, that is, the specific position of electron and the momentum of the electron cannot be determined accurately simultaneously. This scientific law is called Heisenberg Uncertainty Principle. Nature's rules are fixed-strange as it seemed, but that's how it is. These electrons have their own intrinsic behaviour which is steadfast and absolute. Man is just an observer who is incidental to this whole scheme of things.

The electrons occupying the shells in an atom or molecule are continuously in motion. The mechanics of electron movement is nothing like the rules dictating motion of objects which are visible to us. The motion of electron in the limited micro space is totally different and does not follow laws of physics as in Newton's Laws of Motion. Newton's Laws of Motion applies to large bodies, and a good example is the motion of cars and motion of planetary systems. Speed, acceleration, momentum and position can be precisely calculated using Newton's equations. However, Newtonian physics, in the world as we know, breaks down when it comes to movement of electrons in the atom. The physics of it requires a different conceptual understanding. Thinking, that departs from what we are traditionally used to in the macro world.

Allow me, then, to discuss some "bizarre" behaviour of electrons in atoms. You may find it intriguing as it opens the door to the world of quantum mechanics.

As we would understand in classical physics, an electron in an atom occupying an appropriate energy level will be confined to space within the energy barrier. So, under Newtonian laws or classical physics, the electron cannot exist outside the energy level band and will remain within the influence of the atom. The laws of classical physics does not allow the electron to exist outside the influence of the atom. However, in quantum physics, the electron can exist in locations that are totally not predictable under classical physics. Imagine a frog living in a very deep well and has no chance of coming out of the well using its own physical strength but under quantum mechanics the frog has a definitive probability of finding itself up outside the deep well.

The concept of not being able to determine an electron's position and momentum precisely is already difficult enough to grasp but to say that the particle can exist outside the potential energy barrier is something else. In quantum mechanics, the electron has potential to exist outside a potential energy barrier.

It may be strange, but those are the laws of physics. As they always say, truth is stranger than fiction.

Electrons in the atom and molecules have their own rules of physics and they follow the principles of chemistry accordingly, and these rules are the same throughout the Universe and do not change. These scientific laws are absolute.

There is a packing order when it comes to filling up the orbits and atoms strive to achieve stability by attaining the full house in every orbit or shell. The first electron goes to the first orbit, and the second electron goes to the first orbit but with a different spin, the first orbit attains full house and becomes satisfied. The third electron goes to the second orbit till it fills the required 8 electron configuration.

Molecules, which are a combination, of similar atom or dissimilar atoms seek to reach the stability in their electron configuration. Atoms seek partners with its own kind of atom or different elemental atom to achieve stability in the electron configuration, which is 2 electrons in the first shell or 8 electrons in the second shell or 8 in the third shell or 18 in the fourth shell and so forth.

Atoms search for this stability by either giving up electrons or acquiring electrons or sharing electrons to achieve the stable configuration. The adventure in searching for the electrons or giving up electrons to another atom or looking to share electrons with cooperative atoms of same or different kind is the manifestation of chemical reaction between elements forming compounds and molecules. Chemical reaction is the path towards forming stable electron configuration. The behaviour of electrons allows this to happen.

In nature, some atoms would have already attained the stable configuration due to its sequential formation and as such remain in their atomic form. These elements with stable configuration do not search for electrons, and as a result, are inert and are inactive in normal chemical sense. Helium, Argon, Neon, belong to this class of non reactive elements and their outer shell intrinsically have stable configuration and exist as singular atoms.

Sodium and Chlorine offer an outstanding example for exchange of electrons between them.

The total number of electrons in a Sodium atom is 11 and the electrons arrange themselves following the housing rules and end up with the alone electron in the third orbit. To attain a stable configuration, it is easier, for the sodium atom to shed the single electron in the third orbit than to acquire 7 electrons. It has to find another atom that has an electron arrangement that enable it to receive an electron. Sodium should be able to give away the lone electron in its outer shell and the recipient atom should be able to receive the electron. And both in the process achieve stable electron configurations. The Chlorine atom offers such an opportunity to Sodium.

Chlorine has a total of 17 electrons, and the outer orbit carries 7 electrons, short of 1 electron to become stable. Sodium releases one electron and Chlorine accepts the electron from Sodium. Sodium becomes a positively charged moiety with the outer electron orbit now having 8 electrons, and Chlorine atom now becomes a negatively charged moiety with outer stable ring of 8 electrons. Both the entities are stable and are married to each other and are attracted to each other due to the negative and positive charges of their individual moieties.

Either shared or captured or released the final resultant outer shell of 8 electrons makes the end product stable.

All elements react to combine, to form compounds amongst them following the rules of chemistry.

They do this by interchanging electrons or sharing electrons to achieve a stable state. The properties of these compounds are different from the elements from which they are formed. Oxygen, Hydrogen, Gold, Sodium, Nitrogen, Carbon, Chlorine, Sulphur, Iron are all elements. However, water, common salt, methane are all compounds resulting from the combination of different types of atoms.

A water molecule is a combination of two Hydrogen atoms and one Oxygen atom. Common salt is a combination of one Sodium atom and one Chlorine atom. Methane is a combination of one Carbon atom and four Hydrogen atoms. They interplay, amongst themselves, to achieve stability.

The formation of atoms and the arrangements of electrons follow methodical laws, and as such, we can expect regularity amongst the elements. Early scientists did not know much about atom and electron arrangements. However, the regularity of chemical reaction observed led a Russian scientist Dmitri Mendeleev to organise groupings of chemicals (elements) and created a table, placing elements with increasing mass. Elements with similar reactive nature were grouped together. Little did he know that similar electron configuration gave rise to similarities in chemical reactivity. Mendeleev was even able to predict the existence of certain undiscovered elements based on the gaps he noticed in the arrangement. Mendeleev was not aware of electrons and protons and their spatial arrangement in the atomic space during his time. This arrangement of elements led to the creation of the Periodic Table. Many others too contributed to the early groupings of the elements.

PERIODIC TABLE OF ELEMENTS

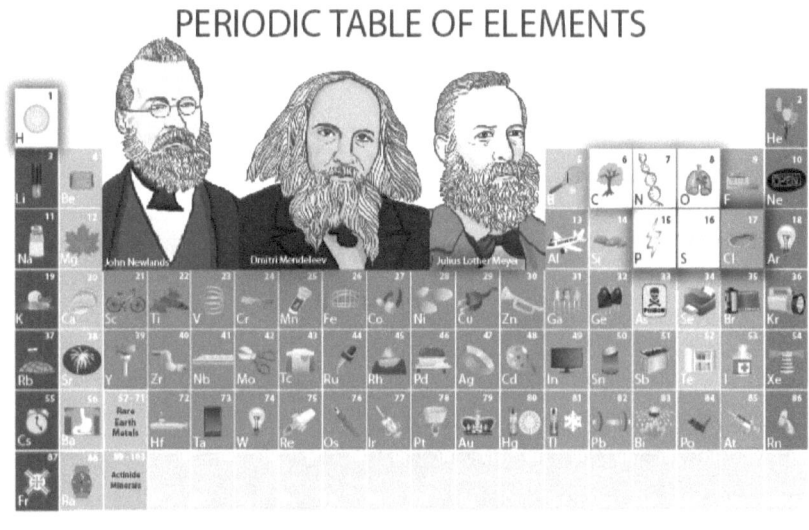

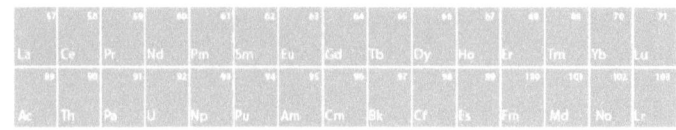

These elements in the periodic table racked havoc on Earth by creating a drama par excellence. They acted out their scenes on stage Earth, performing continuously without break. The Periodic table represents the common family tree of all living things on planet Earth.

But only a few character elements in the periodic table started off the journey initially and let us meet them in the next few chapters.

CHAPTER 4

6 Electron Atom and its Dance

{Grand Entrance of the Character Hero— the 6 Electron-Atom}

Molecules are the basic building blocks of life forms, and one key element was needed to be an anchor and the rallying point to build arrays of molecular structures. This element needed to be flexible in facilitating the creation of multiple molecular forms through multi tasking ability. Where would nature look for such an element?

As elements are methodical permutations of electrons and protons, you can just pick and chose the one that fits the criteria. Nature did pick and chose one on Earth.

The consequence of adding electrons, protons and neutrons in a linear fashion starting with Hydrogen atom will bring us quickly to an atom with six electrons. As we know it now, all atoms move towards stability in the outer shell by having two electrons on the first inner shell and 8 electrons on the second shell. This 6 electron atom's spatial arrangement of electrons according to the chemistry rule of electron housing will have 2 electrons in the inner shell and 4 electrons in the outer shell. The outer ring will be roaring to achieve stability by either acquiring 4 more electrons or by giving away 4 electrons. This atom can give away 4 electrons and become a positive moiety or acquire 4 electrons to become a negative moiety. But giving up 4 electrons is a cumbersome exercise as it is not so easy to find a recipient or donor and the charge on the resultant moieties would be relatively heavy. The electron configuration of this atom with 4 electrons in the outer shell is mid way in forming the stable configuration of 8 electrons. Interestingly, another better alternative, to achieve the same stability, would be sharing of electrons. This sharing of electrons can be done either by pairing electrons with fellow atoms of its' own kind or with electrons of atoms of other elements. The emergence of this atom is a natural progression of filling with electrons, proton and neutron methodically, it was an inevitable consequence. It occurs as the fourth most abundant element in the Universe after Hydrogen, Helium and Oxygen and is eighth most abundant on Earth. We call this element, Carbon.

This element carbon shares the electrons with other fellow carbon atoms and with a host of other elements in its quest to reach the eight electron stable structure Because of its configuration, options to share electrons opened an adventurous path for this element.

This fabulous opportunity of electron sharing gave this atom an open season and a wide avenue to share four electrons with other atoms of its own kind and with atoms of different elements. The ability to form 4 bonds made this element, extremely versatile and nimble. Carbon's configuration gave it the feasibility and flexibility to form numerous bonds with itself and with other elements giving rise to a massive potential of creating enormous range of molecules. The permutation possibility between carbon and atoms of other element is astronomical by statistical standard.

Sharing of electrons is called bonding and a pair of shared electrons, is called a bond. A single bond is where one electron from one atom and one electron from another atom of same or different kind, are shared, a double bond is where 2 electrons from one atom are shared with 2 electrons from another atom or molecule. And a triple bond is when 3 electrons from one atom are shared with 3 electrons from another atom or molecule. Bonds are represented by a line in pictorial form and examples of the bonds between carbon atoms are represented below.

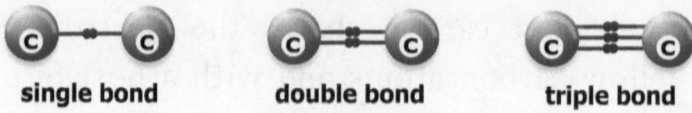

| single bond | double bond | triple bond |

The bond resulting from sharing electrons is also called covalent bonding and this type of bonds result in long chain molecules. In chemistry, elements are represented as symbols in alphabets as in the periodic table, which is a worldwide nomenclature for easy reference and understanding. Each line between the atoms represents 2 electrons in the molecular representation.

The symbol for the entire range of elements can be seen in the periodic table, but for all illustration purposes, the following are some examples:

Symbol H is for Hydrogen, C for Carbon, O for Oxygen, N for Nitrogen and S for Sulphur. A combination of carbon and oxygen gives us carbon monoxide and carbon dioxide, represented in symbols as CO and CO_2. Molecules are also represented as a combination of these symbols as in graphic forms as below:

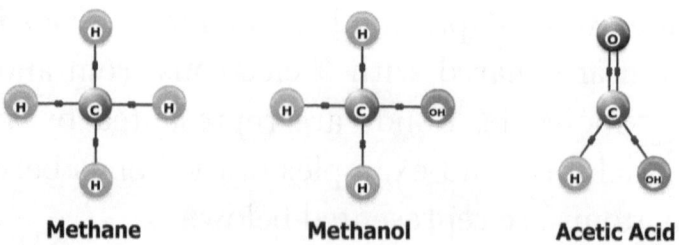

| Methane | Methanol | Acetic Acid |

Examples of molecular structure where carbon formed multiple bonds with itself and with other elements are as below. Carbon element is sometimes represented by 4 spokes of lines indicating the 4 bonds.

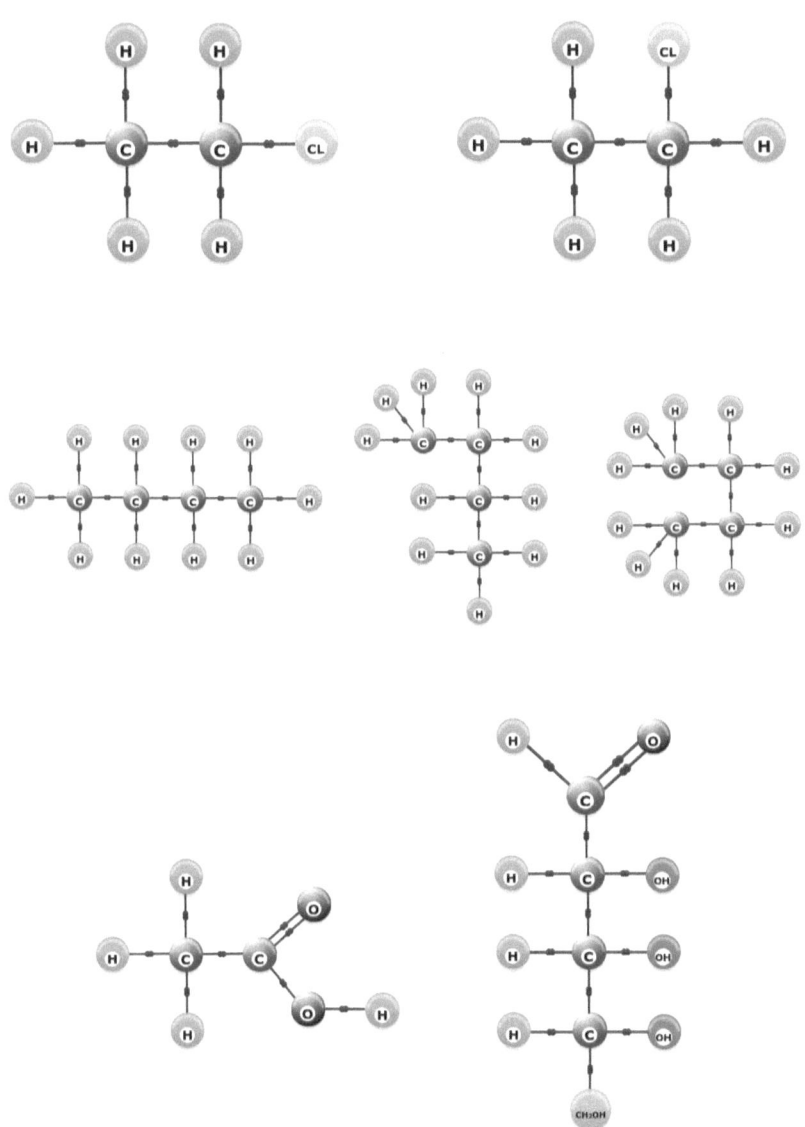

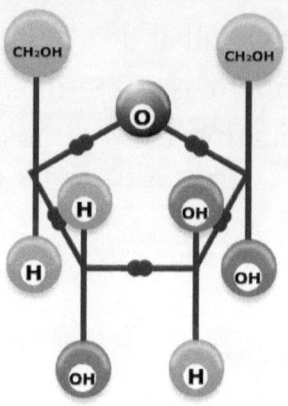

**5 ring cyclic compound of carbon
with participation of hydrogen and oxygen**

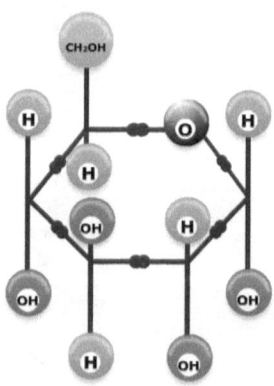

**6 ring cyclic compound of carbon
with hydrogen and oxygen**

These are fairly straightforward sequence of atoms arranging themselves, whichever way they wanted, giving rise to purposeful molecular structures as required by the ecosystem.

It is amazing to see the power of Carbon in organising itself and with other elements. This is elementary for Carbon simply because of the electron arrangement. Nothing is unusual for Carbon to act out with its cohorts to create countless varieties of molecules.

As a result of this possibility of forming multiple bonds, Carbon became central, and the selected prime mover element for interplay with other elements to form molecular varieties according to the dictates of the environment in which it existed. The adventure of Carbon and its other fellow elements in the periodic table resulted in highly complex structural molecules, driven by an ever changing terrestrial environment.

Molecules occur in three dimensional forms and the angle within which the atoms organise and align to each other in relation to one another gives rise to structural forms for molecules, more so when long chains of carbon atoms are stringed together.

Pairs of electrons in nature repel against each other and try to be as far apart as possible from each other. A combination of 1 carbon atom with 4 hydrogen atoms gives methane and structurally it forms a tetrahedron shape to keep all the electron pairs as far as possible from each other.

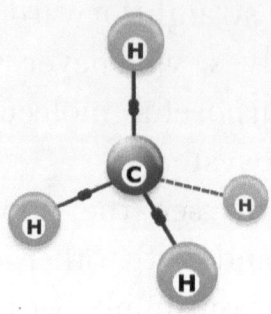

Methane

The hydrogen atoms bonded to the carbon is farthest away from the carbon atom and from each other based on the principle of mutual exclusion, and pairs of electron stay apart from each other. This geometrical shape called tetrahedron is a perfect structural arrangement that keeps the H atoms furthest away from carbon atom and from each other H atoms linked to the central carbon atom.

Basic rule of nature directs chemical bonding and spatial geometrical arrangement, in a mathematical manner. Another ability of carbon equally intriguing is the element's flexibility to form cyclic compounds.

The molecule below is a 6 carbon ring molecule where all the bonds are single bond.

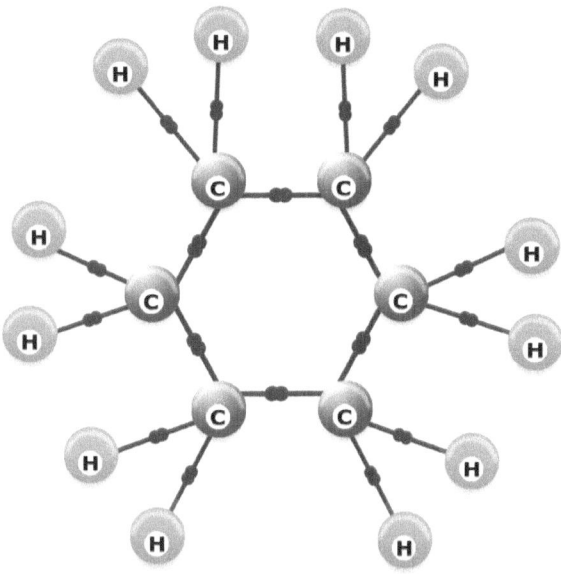

This is called cyclical Hexane

But 2 carbon atoms in the ring can also form double bonds, in which case, the molecule will look like:

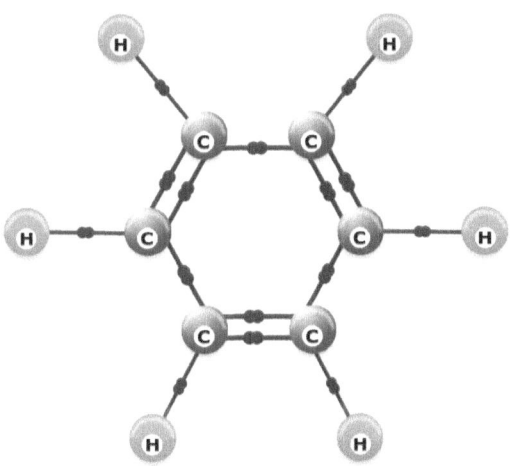

But the double bond position can change in nature as electrons behave to the dictates of quantum mechanics.

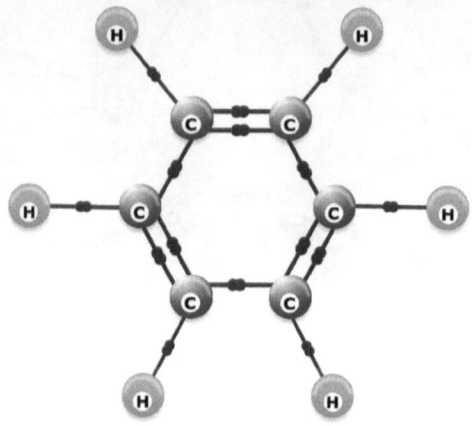

However, the position of the double bonds changes and oscillates between the 2 structural forms rapidly, so fast that each electron from each of the carbon atoms continuously circles around the 6 carbon atoms.

One electron from each of the carbon atom changing position forms a ring of electrons running around the 6 carbon atoms.

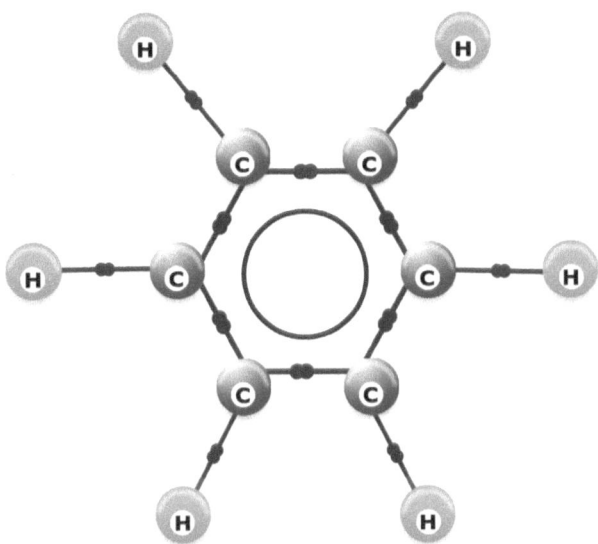

This is the Benzene molecule

The resultant electronic ring around the carbon frame gave rise to a different chemistry from normal cyclic or straight chain carbon molecules, and the chemistry of it came to be known as Aromatic Chemistry.

The limitless probability of forming complex compounds within itself, with hydrogen, nitrogen, oxygen, phosphorous and with many other elements in nature, led to infinite possibilities of molecules exhibiting characteristics which are unique and different.

Carbon is truly the queen of all elements and is a phenomenal atom due to its electron configuration, and its adventure with the rest of its fellow elements is not only mind boggling but extremely awe inspiring, to say the least. Carbon and a couple of other elements in cohorts created and wreaked havoc on Earth. Their behaviour began an adventure beyond imagination of which we are all witnesses of.

But, is there any other element that would have taken the role of Carbon? Possible yes, but on Earth, Carbon was the naturally chosen candidate. Perhaps some other element could have been the rallying point if the environment were radically different on Earth. Maybe some other element is ruling in some other planet or would have reigned or will reign.

If you look at the periodic table, the element below Carbon in the same column and in the 3rd row could be a suitable candidate, as it shares the same configuration as Carbon in the outer shell. That element is Silicon. It will not be surprising to see Silicon based life forms in some distance corner of the Universe.

But on Earth, Carbon's dance with other elements took centre stage and reigned supreme. Since the Carbon march began, the relentless adventure of this element never took a pause.

CHAPTER 5

A Carbon Inspired World

L ife became probable on Earth when macro molecules began to self organise. The self organising property led to construction of molecular systems and structures which also acquired the ability to self replicate and duplicate. Molecules were the raw materials needed to form structural frames to support life forms and to run life processes. The elements and the molecules had it all. Biological processes and mechanisms were put in place alongside the self organising molecules, without which life might not have had a chance to sprout.

The raw materials for all the process of life formation were built from elements, which were carefully picked from the periodic table. Carbon was the central rallying point fully supported by hydrogen, nitrogen, oxygen, sulphur, phosphorus and by a plethora of other elements.

For simplicity, there were three different groups of raw materials used in developing living systems and it can be classified broadly as follows:

(a) Materials for energy storage to run the living system—these are molecules that were used as raw material, to store energy, which could be re used to release energy through biological processes.

(b) Materials for building physical frameworks and for processes for living systems—these are molecules that were used to build biological structure and utilised to aid biological processes.

(c) Materials for information database system— these molecules were used to make templates for assembly of biological structures and to store information.

These groups of raw materials emerged slowly, working interdependently with each other to create an interactive process and system. There is no telling which group of molecules evolved first, but suffice to say, they all emerged in unison.

To fuel the life system, there was a need for an energy source, initially the source would have been inorganic chemicals in early primeval Earth, but over time there was a push to find more efficient ways of energy capture and utilisation.

Identifying the source of energy, capturing energy, storage of energy and utilization of energy were indispensable processes in evolving a living system. Storage of energy was crucial, and all efforts were focussed in developing a process which would store available energy, in physical molecular form, ready for use on demand. The need to store energy drove the system to devise an energy biomass molecule.

Carbon became the base anchor element for the construction of biomass molecules, as it was naturally, a versatile element. Carbon in collaboration with Hydrogen and Oxygen gave rise to numerous energy compounds.

Material for Energy Storage

A group of energy compounds called carbohydrates in the form of glucose, fructose, ribose, sucrose and starch became suppliers of fuels for living systems.

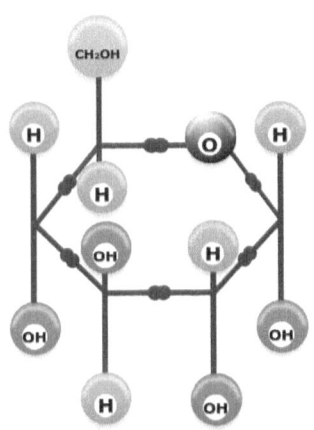

Glucose

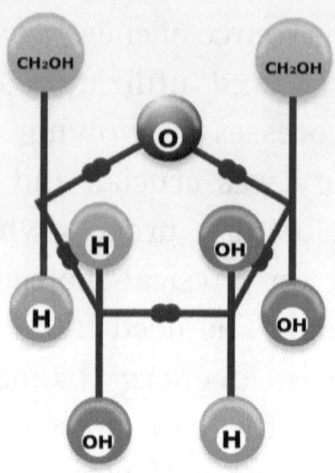

Fructose

The singular unit structures are glucose and fructose. A combination of these units gave complex compounds such as sucrose and starch. This range of molecules called sugars became crucial molecules for capturing energy, acting as energy storage reservoir and for releasing energy for cellular processes when required.

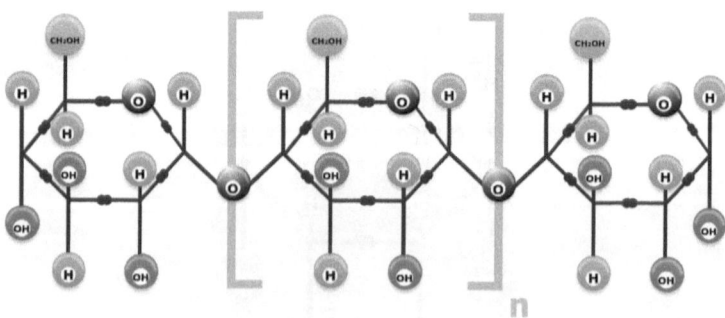

Starch as depicted above is a combination of repeating units of the glucose molecule, which became quite a standard energy storage molecule, in virtually all life forms.

While glucose, fructose, sucrose, starch became energy based compounds, the other analogue became a member of the building block for self replicating molecule. The component member of the replicating molecule is ribose.

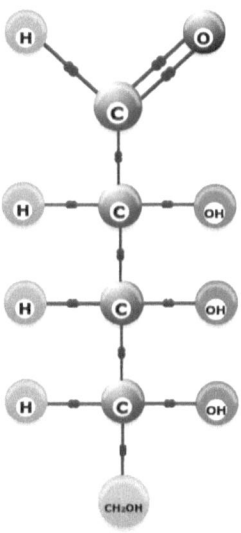

Ribose in straight chain

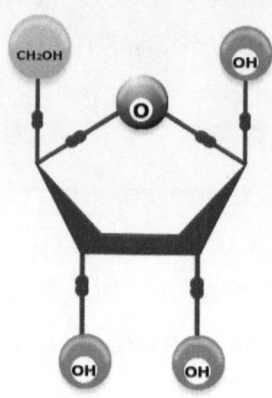

Ribose in cyclical form

The basic building blocks of these energy molecules are atoms of Hydrogen, Carbon and Oxygen, mere 3 different elements.

Given the atmosphere and the right temperature, pressure and the abundance of these elements, formation of these molecules was not only simply feasible but statistically certain.

The Raw Materials for Living Structures and Systems

Hydrogen, Carbon and Oxygen, were the principal players, in organising the molecules we have seen so far. There is another essential element that Carbon is fond of and forms bonds quite easily. This element has 7 electrons in it configuration and sits just beside Carbon in the periodic table. Following the electron housing principle, this element's outer ring has 5 electrons and would need 3 electrons to become stable by sharing electrons and forming 3 covalent bonds. The element is Nitrogen.

Carbon in concert with Nitrogen again incorporating Hydrogen and Oxygen created a series of molecules called amino acids, which became the essential raw material for making living systems.

Amino acids became the components for making biological equipment and were used as building units for structural works and for system process flow within the living mechanism.

Amino Acids have a Nitrogen-Hydrogen group called amine and a Carbon-Oxygen acid group, which can be depicted, as below:

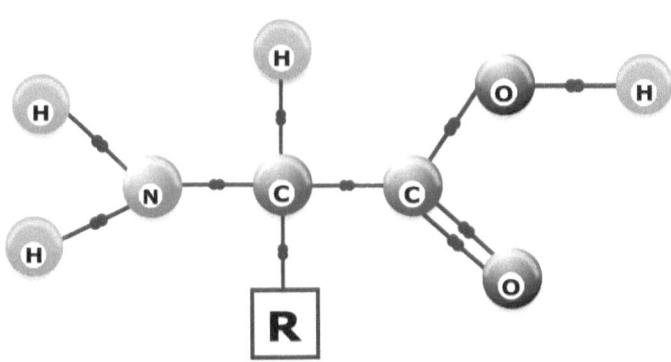

Where, R is a carbon chain composed of carbon and hydrogen.

Amino acids are the very brick and mortar for building protein molecules. The amino acids join through a bond between nitrogen atom on one molecule and carbon atom of another amino acid molecule to form peptides, polypeptides and proteins.

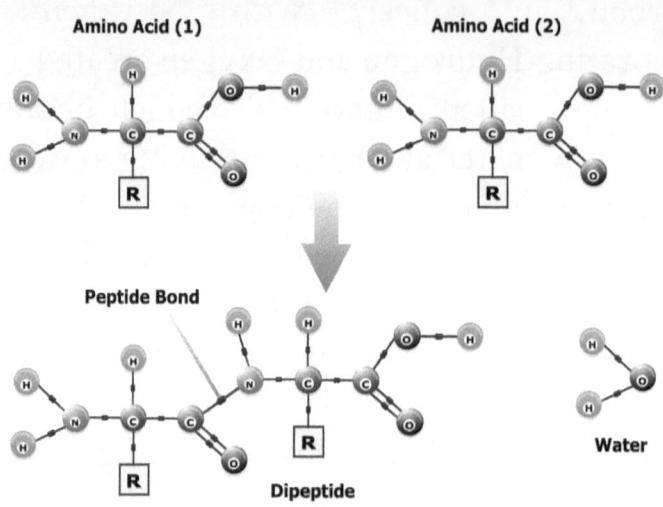

Proteins are the basic material for building structure and framework in living vessels and systems. These proteins are also the very substances that organise reactions and biochemical pathways in the living cells. They act as enzymes in biological processes. Proteins are inherently polymers of amino acids. Living things utilise 20 odd types of amino acids in various permutation and combinations to synthesise different varieties of protein molecules in substance, design and structure.

All biological systems and structures are made up of protein and proteins became frames, piles, slabs that were used in building biological body forms. The chemical properties of amino acids determined the biological activity of the protein molecules.

Some examples of amino acids are:

Nonpolar Amino Acids

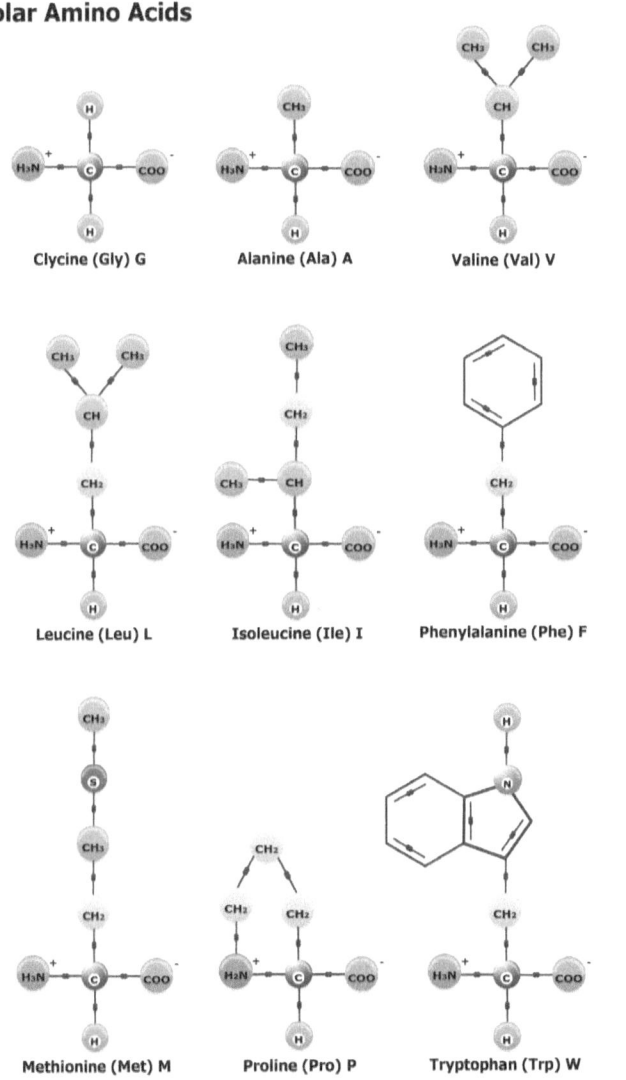

Polar Amino Acids (Neutral)

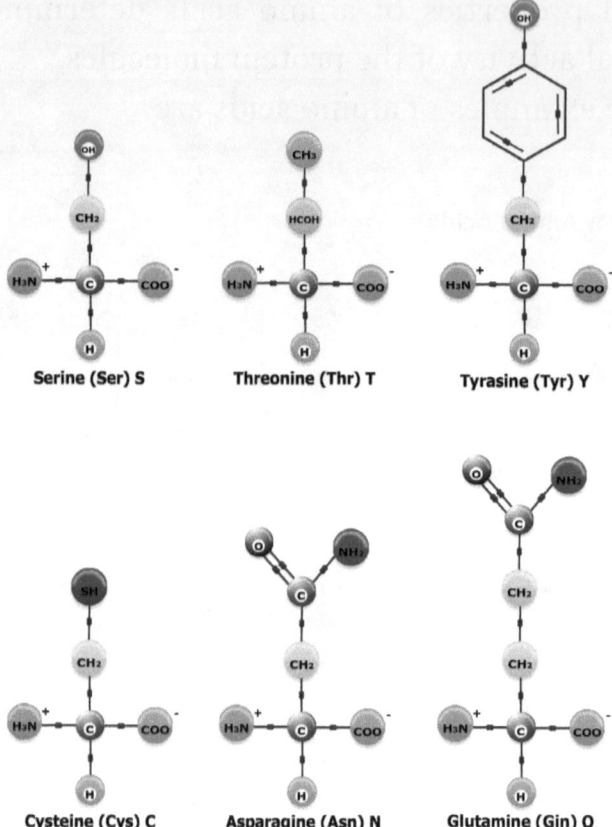

Serine (Ser) S	Threonine (Thr) T	Tyrasine (Tyr) Y

Cysteine (Cys) C	Asparagine (Asn) N	Glutamine (Gln) Q

Acidic Amino Acids

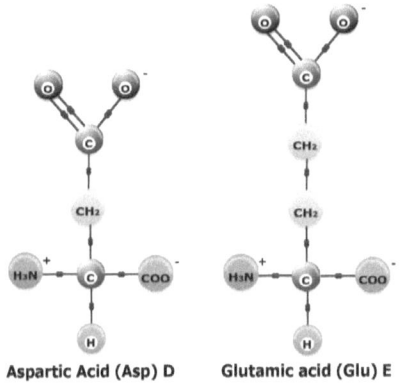

Aspartic Acid (Asp) D   Glutamic acid (Glu) E

Basic Amino Acids

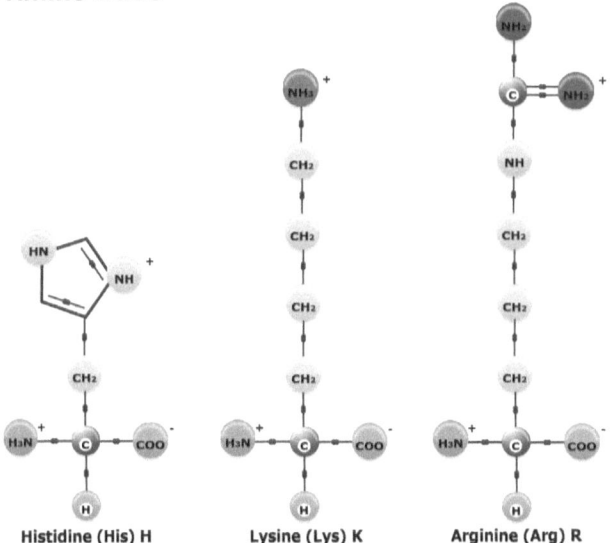

Histidine (His) H   Lysine (Lys) K   Arginine (Arg) R

You will notice the play between Carbon, Hydrogen, Oxygen, Sulphur and Nitrogen in the amino acid structures depicted. They don't look complicated at all.

So, by playing with 20 different types of amino acids in different combination and permutation (similar to building structures using 20 different types of lego units), the living mechanism created shapes and forms that exhibited different characteristics and functions that suited to their requirements under the influence of the environment.

Each sequence of amino acids resulted in unique forms of structures with specific functions. Like lego structures, you can choose what is required and discard the non functional ones. The choice was enormous in front of nature, and nature kept playing with this dough till it got what it wanted.

As in any trial and error method, every time you get some useful results, you would want to safe keep the formula and the sequence. Just as when you cooked up a fabulous meal, you would want to record the recipe and keep it for future reference. You don't have to struggle to remake and retrace the steps each time from ground zero. It would be an excellent idea to record it somewhere so that you can go back to the information file or archive, retrieve it and run the same process. Anyway, how did nature face this challenge of record keeping?

The first step obviously was to make a template that can store the information so that the process can be repeated without having to go through trial and error method every time.

Useful sequences of amino acids needed to be recorded and kept so that appropriate proteins can be made as and when required using the template. However, a reliable and a repeatable manufacturing process also needed to be put in place. What would a sequence planner for production look for? Would a process-driven manufacturing plant with some jigs, energy-generating mechanism with inbuilt information procedures make things run efficiently? Definitely so and such a system, process flow mechanism and standard operating procedures were put in place naturally for the assembly of amino acids.

In the next few chapters, we will look at those molecules that emerged to evolve the system and procedure for the synthesis of protein from amino acids.

"A chain of amino acids"

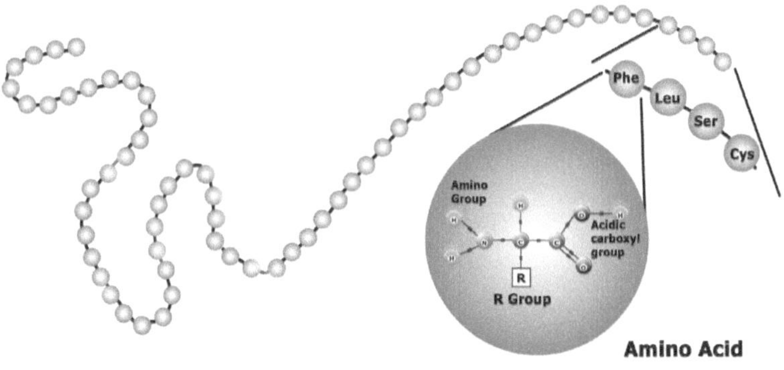

Amino Acid

CHAPTER 6

Concert amongst Close Comrades

A small number of elements worked, in concert, not only to produce energy molecules (carbohydrates) but also produced molecules for processes and structures (amino acids). Apart from Hydrogen which is the basic atom, other elements appear in one segment in the periodic table and are connected to each other, representing a combined ability to cover the entire range of bonding possibilities. Each of these atoms interacted, with each other, to give numerous molecules, which became useful, biological compounds. These biological compounds acted in unison to create a system that gave rise to duplicating and self sustaining entities, which formed the basis of life. The elements themselves are the self generators of molecules and each element played its role singularly and in combination with other fellow elements in the periodic table.

Two other elements closely related to Nitrogen and Oxygen that also played essential roles in life molecules are Phosphorus and Sulphur. Phosphorus apart from forming 3 covalent bonds, like Nitrogen, also has the ability to form 5 covalent bonds. This ability to form 5 bonds landed Phosphorus the role as a bridging element to string together different moieties. We will see Phosphorus' contribution towards stringing frameworks in the replicating molecules, in the next chapter.

Sulphur was incorporated as an important element in 2 of the naturally occurring amino acids. Sulphur and Oxygen exhibit similar elemental behaviour as they both have 6 electrons in the outer rings. Sulphur played its role well in the biology of early Earth when the atmosphere was not filled up with Oxygen. Oxygen proliferated when photosynthesis cropped up, and it did a terrific job of filling up the Earth's atmosphere and also formed the ozone (O_3) layer which protected the Earth from harmful cosmic rays.

Thanks to Oxygen's foray into the atmosphere, otherwise we would be "breathing" sulphur or eating sulphite for all we know, for conversion of energy. But then again, we would have been a different creature all together!

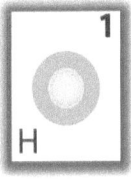

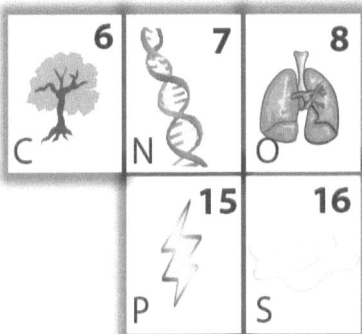

It is not strange at all that Carbon(C), Nitrogen (N), Oxygen (O), Phosphorus (P) and Sulphur(S) are closely related in the periodic table and are kids from the same block. Carbon, Nitrogen and Oxygen, are results of an arithmetic progression series with one electron added on in progression. The difference is in the electron configuration which allowed the elements the flexibility in molecular bonding capability. They were uniquely methodical in their interplay, and they coordinated together well in an ensemble.

Nature chose the simplest of the atoms to start the play of building molecules, to enter into an unchartered adventure and had no specific objective except for just moving forward reacting, with each other, to form molecules, simple and complex.

Carbon is the fourth most abundant element in the Universe after Hydrogen, Helium and Oxygen. Oxygen is the third most abundant element in the Universe. Nitrogen is 78 % of Earth's atmosphere and is the 7th most abundant element in the Universe. That these elements, which were in abundance, could have reacted together naturally at the right opportune temperature and pressure in a cooking pot to produce amino acids, ribose, nitrogen bases and phosphoric acid is not far fledged at all. A very probable possibility!

We have seen energy molecules and amino acid, and it is now time to look at that molecular entity that created the information database system along with the procedural systems that delivered the protein synthesis mechanism.

CHAPTER 7

Evolving the Mother Molecules by Trial & Error Method

Every life form even the simplest of it was driven by self replicating molecules. The self replicating molecules acted as information templates for building biological equipments and structural frameworks. The templates were indeed the flow charts and instructional procedures for the assembly of proteins.

Sculpturing structural molecules for synthesis of protein while in tandem developing a manufacturing process system, was an extraordinarily tedious challenging task, as there were no set processes and everything had to be initiated through trial and error method.

Nature took millions of years to do all sorts of experiments through trial and error to evolve structural living chemicals. A handful of the elements in the periodic table were picked and chosen to develop structural living chemical and biological systems. These elements were in the same boiling kettle.

Molecules' innate ability to organise structural forms like inanimate geometrically perfect crystals is no big wonder in nature. But when structural molecules self replicates through organised systems and processes, then, that becomes the essence of living systems. The evolvement from inanimate chemical substance to living molecules is not a farfetched an idea as elements, by intrinsic nature organise themselves as stable molecules even as an inanimate non living matter. It was obviously a significant step forward from being an inanimate organised structure to a self replicating structural form. And in our case on Earth, these molecules did just that. However, the randomly existing molecules needed to come together to build the jigs and machinery necessary for developing living systems, which included evolving a duplicating process and a record system.

It would be fascinating to look at the jigs and machinery that were naturally devised to systematically synthesise proteins. Nature avoided over engineering by using the jigs as the template for manufacturing proteins, which also doubled up, as the template for self duplicating system. This ensured no lost of information and the same mother design could be duplicated repeatedly.

The building block of the jigs molecule is composed of units of molecular moieties such as ribose sugar structure, a linkage group and a nitrogen base group.

Let us look at the key elements that were instrumental in shaping the jigs molecule. Again, as if, there were insufficient elements to work with, the same four elements Hydrogen, Oxygen, Carbon and Nitrogen, got together to form yet another group of molecules that became structural and directional building blocks for the mother jig replicating molecule.

The building block molecules which are simple construction of polygon shaped structure resulted from a foray of bonding adventure between Carbon, Oxygen, Nitrogen and Hydrogen. These resultant compounds Adenine, Guanine, Cytosine, Thymine and Uracil called Nitrogen bases became component members of the duplicating jigs.

The Nitrogen bases are of two different types, Purines (2—ring compounds) and Pyrimidines (1-ring compounds) have the following structures:

The Purines

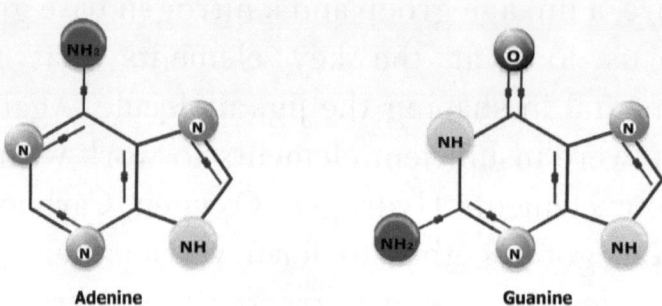

Adenine   Guanine

The Pyrimidines

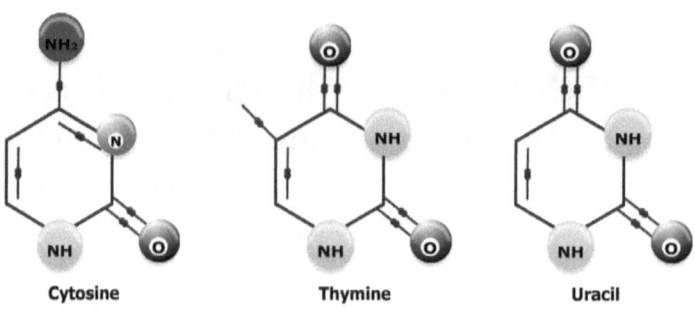

Cytosine   Thymine   Uracil

The molecular jigs needed to have a structural form that was capable of sequencing the raw materials of amino acids into proteins and safe guarding the sequence for repeatability. The molecular jigs were built from units of ribose, nitrogen bases and a linkage group. The singular unit is called nucleotide which became the base repeating unit for the mother jig molecule.

Phosphorus contributed towards building the backbone linkage between the units of nucleotides.

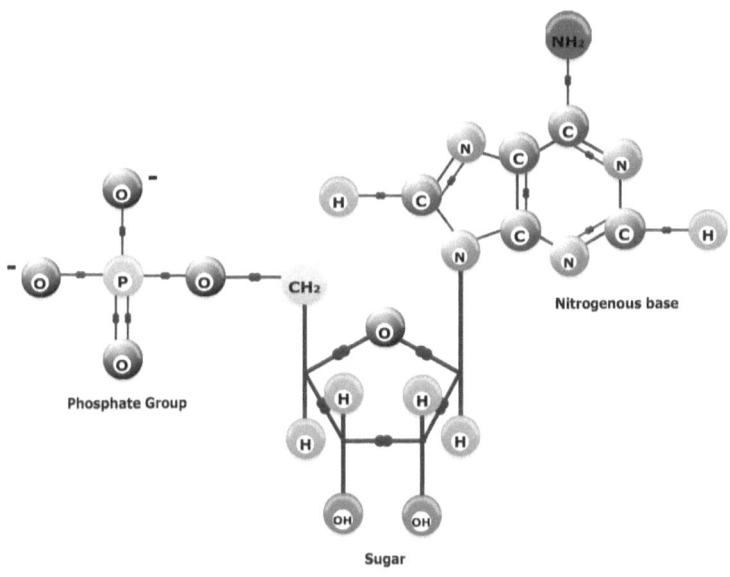

Nucleotide

5 different types of Nitrogen bases, a ribose sugar and a phosphate linkage group started off a journey of no return that brought wonders to the world as we know today. Without them, life could have been nonexistent on Earth. The astonishing feat what elements could do, these molecules did marvellously even better.

Our sojourn into the world of atomic elements is so far confined to 6 elements, Carbon, Hydrogen, Oxygen, Nitrogen, Phosphorus and Sulphur which together delivered a spectrum of interesting molecules.

The play of three elements Carbon, Hydrogen and Oxygen gave us sugars, such as glucose, ribose, fructose, carbohydrates and cellulose.

5 of the elements Carbon, Hydrogen, Oxygen, Nitrogen and Sulphur played together to give us all the amino acids.

Nitrogen also incorporated itself into cyclic compounds with fellow Carbon and Hydrogen to produce Nitrogen bases.

Phosphorus, a close cousin to all the other 5 elements formed a phosphate group with Oxygen to give linkage between one nucleotide to another nucleotide unit.

These nucleotides formed a polymer chains repeating itself with the option to change the nitrogen bases. The ability to form chains of nucleotides opened an infinite series of possibilities. In order to replicate, these chains paired themselves to generate complementary chains, which were capable of independently pairing again giving duplicates of the original strands. The chain building and the pairing abilities made diversity of life forms possible.

Let us look at this fascinating chain formation:

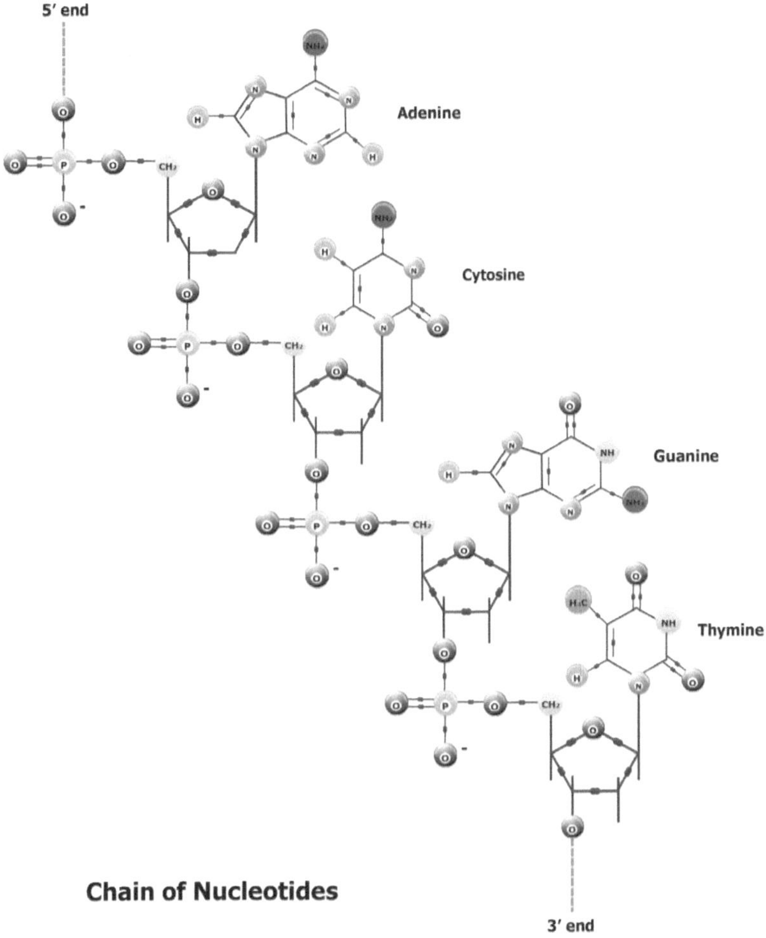

5′ end

Adenine

Cytosine

Guanine

Thymine

Chain of Nucleotides

3′ end

Take note that Thymine and Cytosine are 1-cyclic ring compounds and Adenine and Guanine are 2— ring cyclic compounds.

If you were to pair the nitrogen bases strand to strand and keep the final compound intact and tight, you would have to keep the distance between the 2 strands constant. Looking at the structures of these 4 nitrogen bases, the choice would be to match Adenine to Thymine and match Guanine to Cytosine. Simple lego play! That's exactly what happened!

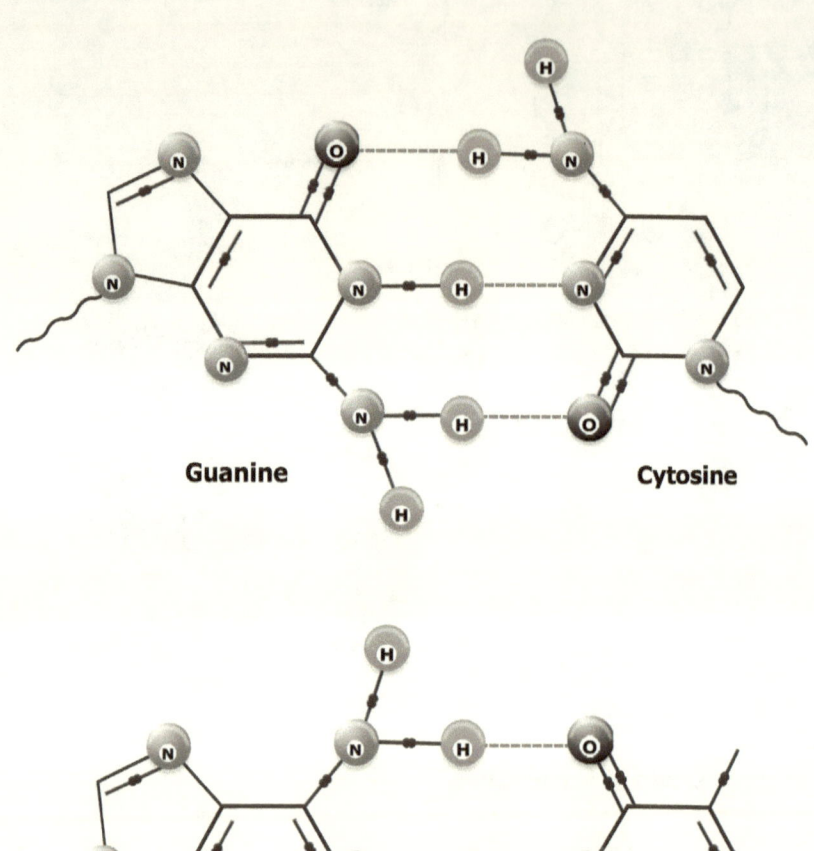

Guanine   Cytosine

Adenine   Thymine

Structurally, if you think like a carpenter and if you needed to fix the 2 strands 3 dimensionally, you will have to keep the strands in the opposite direction to fix it. The final design you would come out with will be a twisted ladder.

So with some geometry and carpentry, 2 strands of nucleotide chain got together by running in the opposite direction by pairing the nitrogen bases—a natural sequence of event. To do this feat, the time taken was no less than 500 million years. If elements can do wonders with their electrons and behave in extremely methodical manners with electrons playing adventurous roles without being directed, the molecules of these sorts were certainly capable of organising themselves to the hilt.

They call this molecule-Deoxyribonucleic Acid or for short DNA—the self replicating molecule!

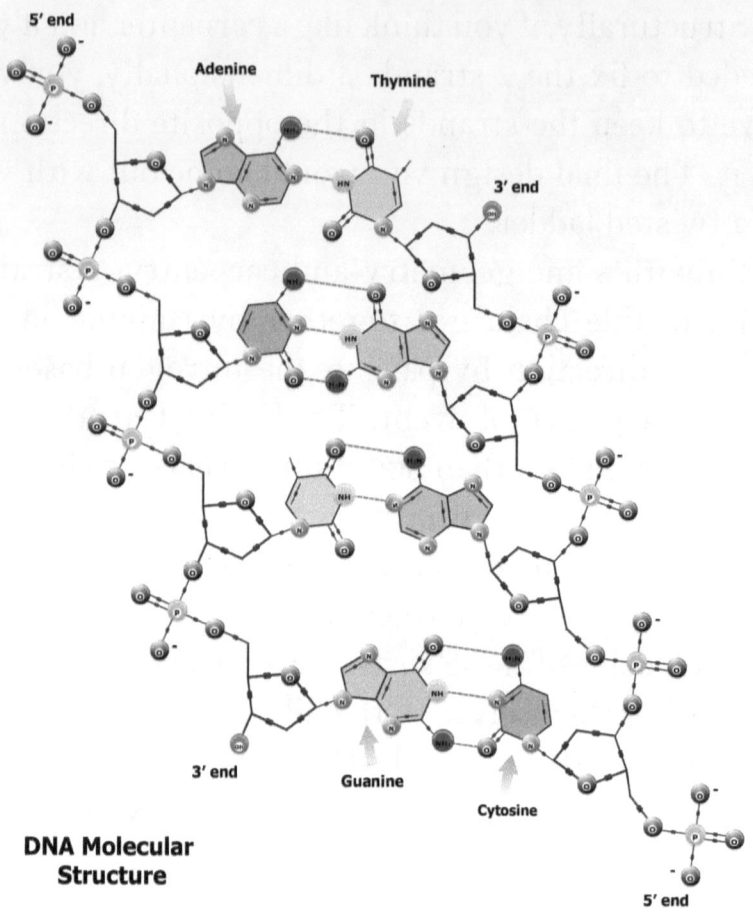

**DNA Molecular
Structure**

The adventure of electron, proton and neutron on Earth took us from elements to molecules and to the replicating mechanism which can be envisaged as:

Electron,Protons,Neutrons

Hydrogen,Oxygen,Nitrogen,Carbon,Phosphorus,Sulphur

Glucose,ribose,amino acids,purines,pyrimidies,phosphoric acid,water

Carbohydrate,Protein,Ribonucleic acid,Deoxyribonucleic acid

Simple replicating mechanism

DRIVEN BY THE ADVENTURE OF ELEMENTS POWERED BY THE INNATE ABILITY OF ELECTRONS

CHAPTER 8

Elementary Play

The molecular structures that we saw in the last few chapters were built from basic chemical components comprising of ordinary atoms and were confined to only Carbon, Oxygen, Nitrogen, Hydrogen, Phosphorus and Sulphur. All these elements are close neighbours in the periodic table and each atom varying in the bond forming ability. Atoms, which were raw materials for the formation of molecules in the living system, were naturally selected due to its electron configurations. In a playroom, given toy bricks of different shapes and sizes, we would have hand—picked these atoms easily to construct the building blocks.

So, there is nothing uncanny or mysterious about the formation of molecular structure that we have so far seen in the previous chapters, and it is not surprising to see these molecules occurring naturally on Earth, given the abundance and availability of the elements comprising them.

There were four groups of compounds that played pivotal roles in making self duplicating molecules.

They are:

The sugar—Ribose

The Amino Acids

The Nitrogen bases—Adenine, Guanine, Cytosine, Thymine, Uracil

And the Phosphate group

We have seen the structure of these above molecules, and how these molecules and polymers were made from molecules that can be stringed up from primary components. The elements were fundamental to the formation of molecules and molecules were playing their own game based on their own rules. Combination and permutation gave rise to possibilities of numerous molecular structural forms, and this led to specific molecules that began their journey to build bigger-than-themselves structures, with no specific goals whatsoever except for its own self.

The basic building blocks selected for the self duplicating adventure were the nucleotides, which became lego type units in the playroom.

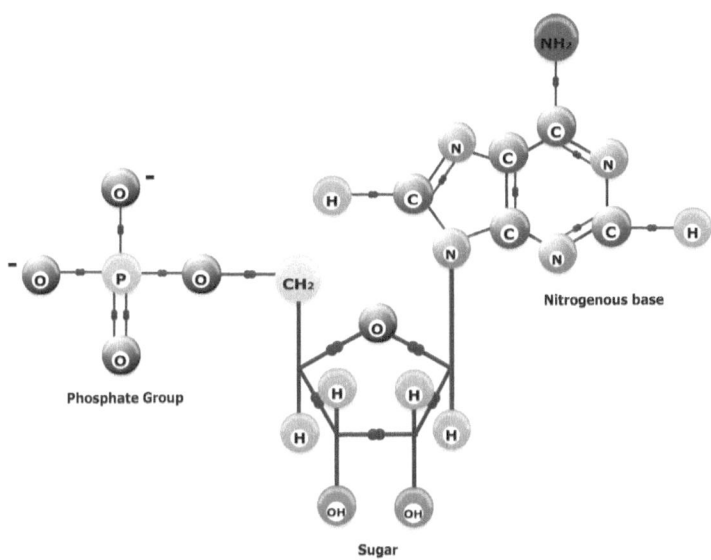

Phosphate Group

Nitrogenous base

Sugar

NUCLEOTIDE

As was explained in the previous chapter, these nucleotide units assembled themselves into long chain strands and organised themselves neatly into double strands. The double stands bridged each other through hydrogen bond, a weak bond, which allows the stands to separate when instructed.

This ensemble of nucleotide units is what DNA is all about!

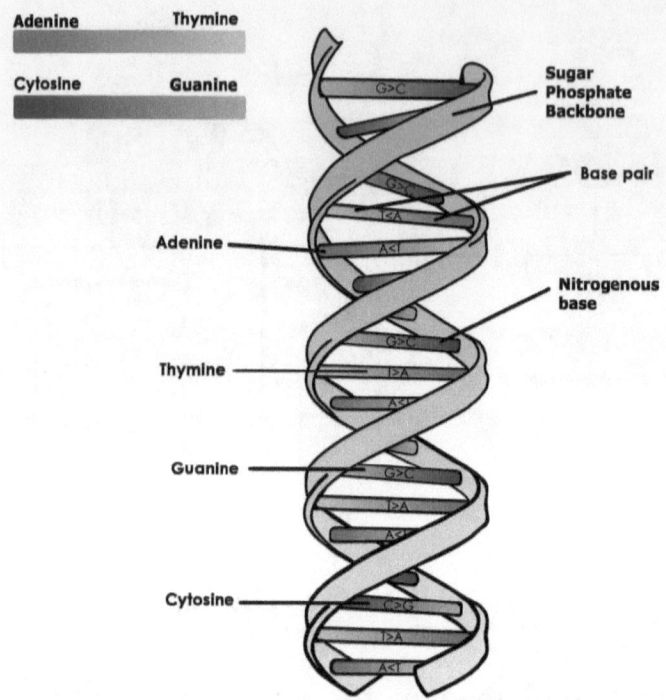

THE DNA MOLECULE

Doesn't DNA molecule look elegantly straightforward? It should, by now. We have spent a lot of time to arrive at this molecule, beginning only with a couple of different elemental atoms. You should be able to recognise all these elements in the DNA. The elements you see are basic ones anyway, only 6 elements, out of 92 naturally occurring elements. Less than 7 % of the various elements found on Earth constituted DNA elemental makeup. That's smartest optimal use of available resources. The molecules were self selecting and were choosing the right type of structures, driven by an expansionary urge to build.

However, how long do you think it would have taken to make this selection?

I speculate, the elements and molecules took 400 to 500 million years to get the primordial precursor of these molecules in place. When the environmental conditions were suitable, the elements and molecules acted together through computational trials. And they had all the time to do the selection of the appropriate molecules. But then the molecules had no planned path and no glimpse of the future.

Based on the knowledge we have today, my speculation is, the molecular play would have happened about 4~3.5 billion years ago on planet Earth. The environment, then, on Earth was remarkably different. The Earth was then hydrogen rich, and hydrogen played a key role in all the organic molecules. Geothermal power source during the cooling phase of the Early Earth fuelled the molecular building exercise. The atmosphere then in the early Earth was extremely different from the likes of today's relatively calm blue sky.

The molecules had a long period of time to do all the exhaustive, countless experiments and trials. During these trials, it selected the functional sequences and discarded those that were not so useful. But the molecules kept doing those experiments incessantly and moved forward in small steps to a purpose for itself.

We all know that the DNA did happen, the atoms and molecules interacted. But how and what made it possible?

To answer this, we have to get back to nature's test tubes and crucibles.

In any case, what was the game plan of these molecules?

CHAPTER 9

Water, Water Everywhere

Nature's crucible that made biological molecules possible in the early Earth needed many support systems for it to emerge. A medium of connectivity was necessary for the elements and molecules to intermingle to facilitate biological formations. Water, a chemical combination of Hydrogen and Oxygen, exist as a liquid, gas and solid within Earth's atmospheric pressure and temperature today and existed as a liquid and as steam some 4 billion years ago during the formative years of the young planet.

The fact that water existed as a liquid under Earth's atmospheric and gravity was crucial in providing the condition for the elements and molecules to react. Water is the primary motivator molecule of life on Earth, without which, those replicating life molecules we talked about could not have come into existence and performed the way they finally displayed.

The characteristic of the water molecule stems for its structure and the spatial sharing of electrons between hydrogen and oxygen. The water molecule is composed of 2 Hydrogen atoms and 1 Oxygen atom—the 2 Hydrogen atoms share their 2 electrons with one atom of Oxygen. All three atoms attain the required stable configuration through this molecular arrangement.

In spatial orientation, pairs of electrons will keep away as far as possible from other pairs of electrons. The tetrahedron structure is best suited for all the electron pairs to be furthest away from each other. However, the 2 pairs of non bonding electrons in the Oxygen atom geometrically push the hydrogen moieties closer, making the water molecule bipolar.

Water Molecule

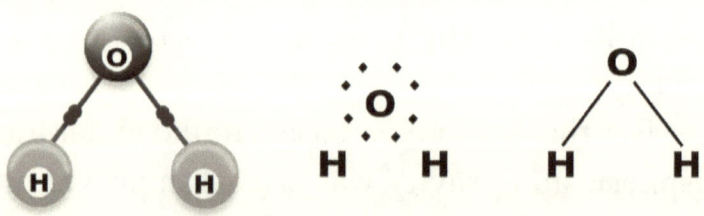

The polarity in the molecule also allowed water molecules to link to each other through hydrogen bonding.

The structural orientation of the water molecules and the polarity of the water molecule gave it the ability to dissolve majority of chemicals. This created the conducive medium for molecules to come together physically to interact chemically.

A system evolved where medium of water became the standard for all molecules to rally around to work and play together. The cooking pot was full of water. The pot created an enclosure for the molecules to reside within the boundary which made further reactions possible for the molecules to advance toward creating the replicating entities. Without water, there might not have been any play for these molecules on Earth. The adventures of elements were made possible and easier, by this marvellous molecule H_2O.

CHAPTER 10

Nucleotide-Amino Acid Venture

The watery crucible where self replicating and process molecules were housed needed to be enclosed to shutout outside influence. Proteins came to the aid to give the structural, protective envelope for the enclosure, with the help of another related compound called lipids. This self contained enclosed capsule developed into the biological cell.

Proteins were also involved in regulating reactions within the enclosed cell. A system evolved to become a self contained entity which also kept the information data bank intact. Information, which was found to be useful for carrying out the functions of the duplicating system, was coded and kept in the databank. The databank molecules were able to duplicate themselves. Having a double strand made the duplication process efficient. When required the double strand could split and build a strand complementary to it. Duplication became the foremost criterion for living systems in its quest for self propagation.

The evolving duplicating living system was driven by a protein manufacturing process. The type of protein manufactured was determined by the sequence of amino acids it was made up. The information on the sequence of amino acids for manufacturing different types of proteins was kept intact, while the system itself was evolving.

The system of identifying amino acid and storing information for sequencing the amino acids into various protein types happened simultaneously and in tandem. The mechanism linked the very system of identifying the amino acid to the sequencing template. If a colour red was used to identify a particular amino acid, then the colour red appeared in the sequencing template, whenever, the particular amino acid was required.

Nitrogen bases, which formed the basis for the DNA molecules, were also naturally selected for identifying amino acids. The sequence of the nitrogen bases in the DNA molecule determined the sequence needed to form protein molecules and the code for amino acids for the synthesis of protein resided in the same place, in the rungs of the DNA molecules. The amino acids, the nitrogen bases and the structure of DNA molecule evolved concurrently and simultaneously.

Though thousands of a variety of amino acid can occur theoretically, only 20 standard amino acids were chosen in nature for coding. Out of these 20 amino acids, thousands of proteins could be designed. There was no further need to select more than what was sufficient. Optimisation of resources permeated the march towards the evolving living system. Limiting the number of amino acid for permutation for making of protein was part of this optimisation. It was much easier to cope with less variable factors especially when it came to data storage and retrieval.

In order to identify 20 amino acids, we would need 20 different codes. Nature ruled that the best way forward is to minimise the number of codes and make the system less cumbersome. The coding and information storing system evolved to fulfil the following conditions:

(1) The ability to duplicate and self replicate the codes.
(2) The ability to code for not less than 20 amino acids.
(3) Minimise number of coding permutation and number of coding unit.

The codes for amino acids have to occur in complementary pairs, to ensure duplication of the codes. Therefore, the system demanded even number of the code bases, i.e. 10, 8, 6, 4 or 2, for duplication to be viable. The coding system must be sufficient to code for the 20 amino acid or more. Again a minimalistic approach was adopted.

Assuming only 2 nitrogen bases were selected for coding, then, the system can uniquely identifying only 4 amino acids (2 X2).

If 3 nitrogen bases were selected for coding, then, the system would allow 27 different amino acids (3 X 3X 3) to be identified.

However, if, 4 nitrogen bases were selected, then we will have a system code for 256 amino acids (4X4X4X4).

As there was only a need to code for 20 amino acids, 3 bases were sufficient enough. Since an even number of nitrogen bases was required for complementary pairing, then, the minimum required would be 4 nitrogen bases. There are, however, 5 naturally available nitrogen bases, Adenine (A), Guanine (G), Thymine (T), Cytosine(C) and Uracil (U).

The requirement was for only 4 bases and hence DNA, the information data warehouse, selected 4 nitrogen bases, Adenine (A), Guanine (G), Thymine (T) and Cytosine(C). DNA's counterpart in the assembly of amino acids called Ribonucleic Acid (RNA) also chose 4 nitrogen bases, Adenine, Guanine, Cytosine and Uracil (U). The difference is in Uracil which occurs in RNA in place of Thymine in DNA. However, note that Thymine and Uracil have same paring properties. Adenine will match only with Thymine or Uracil. Guanine will pair only with Cytosine.

Base on the minimalistic approach model, nature chose a 3 member nitrogen base system from 4 different types of nitrogen bases, as code for the amino acids. Adenine, Guanine, Cytosine and Uracil, were the chosen ones. There could have been other options but natural selection preferred the 4 bases for pairing and 3 base options for coding—good mathematics!

The 3 base code system for coding using 4 available bases for pairing gave us 64 (4 X 4 X 4) different codes.

Adenine-Adenine-Adenine (AAA), Adenine-Guanine-Adenine (AGA), Adenine-Cytosine-Cytosine (ACC), Adenine-Uracil-Uracil (AUU) or Guanine-Cytosine-Cytosine (GCC), Cytosine-Uracil-Guanine (CUG), Uracil-Adenine-Cytosine (UAC) are some of the codes that can be computed. Since only 20 amino acids need to be coded, 64 codes were more than sufficient, and therefore, more than one code coded for the same amino acid. Some codes did not code for any amino acid and were used as start or break signal in the manufacturing process. In protein synthesis, the coding sequence in DNA is translated to a RNA strand called the messenger RNA or mRNA, called codon. Another RNA molecule called transfer RNA or tRNA, the anti codon, carried with it specific amino acid according to the codon. The codon became the template to assemble the amino acids in sequence, as dictated by the part of the DNA strand. The amino acids were assembled according to the nitrogen base sequence in the DNA molecules through the complimentary codon.

CODON TABLE OF NITROGEN BASES CORRESPONDING TO AMINO ACIDS

CUU	(Leu/L)Leucine	CCU	(Pro/P)Proline	CAU	(His/H)Histtidine	CGU	(Arg/R)Arginine
CUC	(Leu/L)Leucine	CCC	(Pro/P)Proline	CAC	(His/H)Histtidine	CGC	(Arg/R)Arginine
CUA	(Leu/L)Leucine	CCA	(Pro/P)Proline	CAA	(Gln/G)Glutamine	CGA	(Arg/R)Arginine
CUG	(Leu/L)Leucine	CCG	(Pro/P)Proline	CAG	(Gln/G)Glutamine	CGG	(Arg/R)Arginine
AUU	(Ile/i)Isoleucine	ACU	(Thr/T)Threonine	AAU	(Asn/A)Aspragine	AGU	(Ser/S)Serine
AUC	(Ile/i)Isoleucine	ACC	(Thr/T)Threonine	AAC	(Asn/A)Aspragine	AGC	(Ser/S)Serine
AUA	(Ile/i)Isoleucine	ACA	(Thr/T)Threonine	AAA	(Lys/K)Lysine	AGA	(Arg/R)Arginine
AUG	(Met/M)Methionine	ACG	(Thr/T)Threonine	AAG	(Lys/K)Lysine	AGG	(Arg/R)Arginine
GUU	(Val/V)Valine	GCU	(Ala/A)Alanine	GAU	(Asp/D)Aspartic acid	GGU	(Gly/G)Glycine
GUC	(Val/V)Valine	GCC	(Ala/A)Alanine	GAC	(Asp/D)Aspartic acid	GGC	(Gly/G)Glycine
GUA	(Val/V)Valine	GCA	(Ala/A)Alanine	GAA	(Glu/E)Glutamic acid	GGA	(Gly/G)Glycine
GUG	(Val/V)Valine	GCG	(Ala/A)Alanine	GAG	(Glu/E)Glutamic acid	GGG	(Gly/G)Glycine

It is a simple manufacturing process using template (DNA), codon (mRNA), anti codon (tRNA) to transfer, decipher and link the amino acid in a sequence to manufacture protein. Information contained in the DNA molecule is retrieved and passed on to a template mRNA that codes for the amino acid. The tRNA brings along the corresponding amino acid and links the amino acid through peptide bonds guided by codon of mRNA and anti codon of tRNA.

The DNA is the information warehouse data bank for the protein manufacturing process. The translation from the DNA template to strings of amino acid can be depicted as:

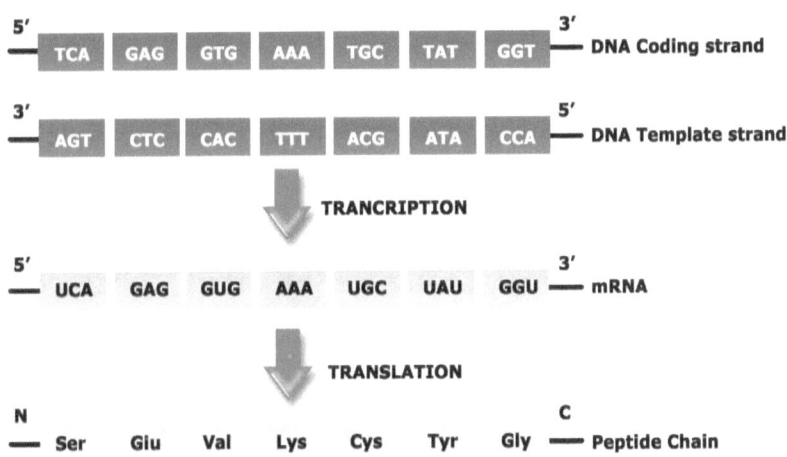

The protein manufacturing process can be pictorially represented as below:

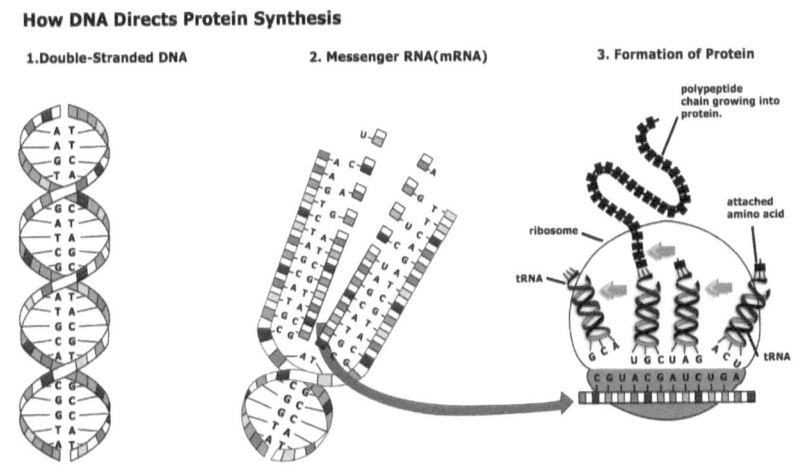

How DNA Directs Protein Synthesis

1.Double-Stranded DNA   2. Messenger RNA(mRNA)   3. Formation of Protein

The technicality of this process is complex, no doubt, but conceptually the mechanism is fairly unsophisticated and straight forward.

This system took more than 500 million years to evolve and that too rather slowly and painstakingly using whatever available elements and compounds in an environment where trials were carried out by precursor molecules. On achieving stability, there was no stopping of this march of DNA and it went on continuously experimenting, changing and developing varying sequence of amino acid codes befitting the environment. It was like a production house running amok.

The amino acid permutations using 20 amino acids opened limitless opportunities to produce proteins with varying properties. The system that emerged inadvertently led to an enormous theoretical possibility of creating varieties of protein structures. This further opened up yet another adventure of sort using protein as raw material to mould living structures.

Simple life forms began to flourish about 3 billion years ago and continued the journey into developing different types of flora and fauna. Diverse forms of species sprang up along the way, some survived, but many went extinct as a result of natural and astronomical catastrophic incidents.

But one thing that remained constant was the system. The template, the nucleotide coding system, the amino acids for making proteins, never changed over billions of years. The same system was used for duplication by all living forms that appeared on Earth. Plants, algae, grass, rain tree, red tree, bacteria, worms, virus, elephant, dinosaur, lizards, kangaroo, jelly fish, sharks, whales, orang-utan, seahorse, various Homo species, daffodils and mudskippers all share the same biological amino acid coding system for the manufacture of protein and we all use the same platform, mechanism and template. The system never changed, the coding system remained unchanged throughout the entire living kingdom on Earth. All living matter and things shared the same platform inherited from the primordial experiment. This allowed the life forms to coexist and use each other as resources for common amino acids and re manufacture required proteins using the same platform, the DNA-RNA mechanism.

We, the plants and animals, are close relatives to one another, as we all share the same mother molecule that was responsible for all life forms on Earth. Without the advent of amino acids, coding system, DNA and RNA, life forms would not have flourished on planet Earth. These chemicals and molecules co existed together and developed to what was to become the singular dominant platform for all living beings to exist. Man was an incidental occurrence in this protein march, and as it had it, the evolved intelligence in *Homo sapiens* made us more than a mere observer in the grandiose adventure of the elements and molecules.

The protein synthesis system propelled emergence of all sorts of species.

It appears to be a mystery as to how these molecules would have come together, and I don't think we are very far away from discovering it.

But there is nothing to stop us from speculating and we shall.

PART TWO

THE SUCCESSFUL TRIAL THAT MADE LIFE POSSIBLE AND THE WAY BEYOND BIRTHPLACE EARTH

CHAPTER 11

Millions of Crucibles to Experiment

E arly Earth was able to nurture molecular formations that eventually led to macro molecules, which were able to, self duplicate and replicate.

The fact that amino acids, sugars, nitrogen bases, could be formed, when Carbon, Hydrogen, Oxygen, Nitrogen, Sulphur, Phosphorus came together is no big mystery. However, the formation of these molecules need not have necessarily started from elemental origin. Simple compounds such as methane (CH_4), water (H_2O) or ammonia (NH_3) could have been the precursors. Chances would have been greater for the molecules to interact to form amino acids or RNA or DNA in an aqueous medium in semi solid phase. The watery world provided the right environment for the co mingling.

We know it happened, but the question is how did it happen?

We may not be dead sure as to which came first, the DNA or RNA or amino acids or proteins or the enzymes. Given the co-existence of these molecules together with the associated systems, structures and mechanism, one will be led to conjecture that these molecules evolved together feeding each other's needs, domesticating each other in co-evolving the mechanism and the system in unison. The molecules could have gone through numerous stages of change and development, one acting as the precursor for the other.

Research is unceasingly going on to find out the origins of these molecules, and we are not too far away before we get to the bottom of this very soon. After all, nature took a very long time to build, so creatures of nature like us will certainly take some length of time to discover the grand beginnings of these molecules. However, there is nothing to stop us from doing some speculations. I believe the nitrogen bases, the sugar ribose, phosphoric acid and the amino acids were the participants in the primordial colloidal medium within the watery ocean of early Earth. And in some corners of that aqua ocean, there existed semi solid paste of earthly substance where the interplay of amino acids and nitrogen base molecules was made possible.

The nitrogen bases and amino acids aligned in complementary form while ribose and phosphoric acid groups were attaching themselves as a backbone to support the nitrogen bases (after all, it is no absolute wonder for inanimate crystals to juxtaposition itself in perfect harmony in crystals formation). Millions of experimental stations would have existed in the wilderness of early Earth. Each one of the crucibles would have had the potential to make precursor molecular structures. Possibly there were millions of failures before the molecules hit some milestone.

The semi solid medium and the water molecules would have formed flexible structural grid for the reactions to take place. This medium would have kept the molecules in place to build an assembly line for the emergence of amino acid chain and the growth of nitrogen bases stringed with ribose and phosphoric group. While amino acids were forming chains through peptide bonds, nitrogen bases alongside were forming their own chains supported by ribose and phosphoric acid. And this build up of amino acids and nitrogen bases would have happened simultaneously, adjusting to each other in terms of molecular space and spatial arrangement to arrive at a co existing molecular system and process.

Molecules and bonded atoms organise themselves three dimensionally in spatial form to fulfil the atomic arrangement.

When elements or molecules react, the resultant molecules can occur in two different spatial forms, one a mirror image of other but both forms exhibit the same chemical properties. These mirror image molecules cannot juxtapose. It's like a right hand and left hand, they are mirror images, but they are not identical. In some symmetrical molecules, there is no difference between the mirror images.

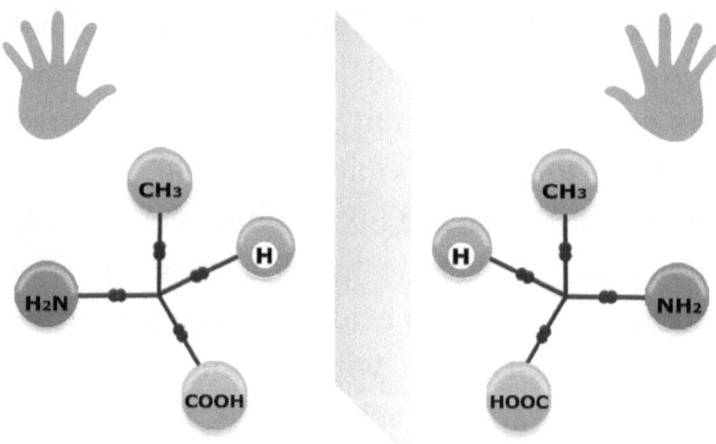

The amino acids above have the same chemical properties, but they are not the same. They are mirror images.

Amino acids and ribose sugars have different mirror images. Biological systems cannot be based on a mixture of mirror images of the same chemical molecule. It is like putting square blocks in round holes. We cannot build a house using different interlocking brick designs. Hence there was no choice but to keep only one type of mirror image as the building block and reject the another mirror image, as picking both types at random could have been a disaster.

Nature's procurement of raw material was stringent and extremely selective. Nature picked only the relevant sided image molecules for the assembly of polymer molecules. Some sort of scaffolding structure to hold the molecules could have aided the process. Medium of semi solid substance such as colloidal watery clay would have been a good scaffolding platform for the crucibles to build the nitrogen base-ribose-phosphoric group backbone framework alongside the amino acid polymer sequence.

The building of the back bone strand and sequencing of amino acids required selection of the appropriate handed molecules so that the structure of the strand and polypeptide were consistent and uniform. Once chosen, there was no turning back.

The option to pick right handed or left handed molecule was fifty-fifty and all it mattered was which mirror image first successfully inter played in the early earthy pot. Once successful the same mirror imaged molecules were chosen consistently without exception. Uniformity and Standardisation!

The amino acids in the polypeptide sequence were build entirely on left handed amino acid moieties and the sugar ribose was based on right handed mirror molecules. And this is almost uniform in all of the living things on Earth. What led to the selection of left or right handed moieties was not of any consequence. But subsequent selection of amino acids or ribose that was added on just followed suit aligning itself to the similar handed molecule type in developing the framework set in the colloidal crucible pot.

The nitrogen base-ribose-phosphoric acid units were painstakingly assembled unit by unit in a repetitive manner to make a single line strand while the other playmate amino acids were aligning itself alongside the single line strand. Surely, an extremely slow trial and error process!

Choosing the right amino acid from a pool of 20 odd types was paramount. This would have dictated the choice of nitrogen bases amongst 5 different types. Both the nucleotide and amino acid were at play compromising and selecting the appropriate amino acids, changing or amending as required to ensure the resultant peptide molecule was of some use in its structural form. The strand of nitrogen bases with ribose and phosphorus group unit kept duplicating itself and adjoining it to its own self as in polymerisation while corresponding amino acid and appropriate nitrogen base sequence were simultaneously assembling itself.

The adventure continued with the formation of RNA and DNA, after having developed the single strand moiety. These molecules replicated and duplicated for the sake of molecular adventure-a self organising property. There was no predetermination, and there was no road map or goal. It was a free flow adventure for these molecules, and it was totally ad hoc-a limitless world with no rules. No game plan.

Only one trial from millions of crucibles emerged successfully, and it remained unchanged, and all other millions of experiments were either totally discarded or never emerged. Only one system prevailed, and it totally dominated the Earth.

DNA-RNA with the amino acids, the coding system and the associated processes emerged as the singular successful system from millions of the trial and error crucibles that would have existed during early Earth.

And we owe our beginning to that one successful crucible.

CHAPTER 12

Ribonucleic Acid (RNA) and its Cohorts

Primordial nucleotide chain could have led to Ribonucleic Acid, (RNA) and that moiety together with amino acids managed a simple duplication and the protein building process. An initial single strand of a couple of codes and amino acids was all it required to kick a key growth process. The combination of a nitrogen base with ribose coupled with the phosphorus group formed a unit for repetition. This unit kept duplicating while changing the nitrogen bases. Simultaneously, corresponding amino acids aligned alongside based on the 3 base code.

The ability to grow the nitrogen base-ribose-phosphorus group (the nucleotide) as polymer while changing the nitrogen base was a crucial breakthrough in coding for amino acid sequence. This interaction between the nucleotides and amino acids eventually led to the growth of the nucleotide chain. The single chain nucleotide polymer became the basis for Ribonucleic Acid (RNA).

A change in one nucleotide leads to a change in the amino acid sequence. A single change profoundly changes the character of the protein produced. The primordial RNA would have interplayed with the amino acids to build itself into a meaningful sequence template. The relentless experiments of the four different types of nucleotides in the formation of various amino acid chains began a long, sprawling adventure of nucleotides inspired family. The nucleotides and amino acids would have worked together feeding each other towards synthesising the required proteins.

Once the duplication processes and assembly lines were ready, they were set to conquer the environment while their arsenals were adjusting themselves to the demands of the changing world around it. The primordial RNA molecules and their macro complex off springs furthered their adventure into domains which were unchartered with no pre planned pathway. The adventure was totally filled with uncertainty.

The RNA molecules were activity driven armed with the ability to duplicate itself. Having engaged amino acids and the mechanism for building proteins, their appetite for further growth and advancement was unstoppable, not even the environment, however, harsh, was any deterrent. The molecules were set in their own path of fresh, unique adventure. It acted out thoroughly in a wild manner to its own purposes, while improving the art of duplication and propagation.

The RNA was a working molecule, and it developed an array of soldiers. Steadfast, but modifying now and then. RNA molecules played the role of a leader and doubled up as working members of ground active combating soldiers. However, there was a need for a supreme general to dictate the movement of the army. This became imperative during the course of events to develop a more sophisticated information flow system that could enable the whole system to be more efficient and effective in moving towards a better structured mechanism with the aim of self preservation. The supreme molecular structure needed to be similar to RNA but with a focus function of evolving a permanent, safe depository of information that could be relied upon and called upon as and when needed, while the RNA functioned as worker molecule. The aim was to develop a molecule that had the property to perform duplication with the ability to protect and manoeuvre itself to accommodate millions of units of information packets. The information bits were the nucleotides sequences.

This information depository molecule is your deoxyribonucleic acid, DNA, an off shoot of primordial RNA. As we have seen earlier, DNA occurs as a double strand helix, one strand running in the opposite direction against the other. Both strands of DNA are strings of amino acids codes for protein synthesis. The complementary strands safeguards the code and keeps the sequence intact and the strands can separate and can arrange themselves to replicate another complementary strand alongside. This ensured repeatability, while carrying along with it any alterations in nucleotide bases. The DNA molecules are a bunch of self preserving self replicating entities.

The matching of the bases between the 2 different strands running seamlessly in 3 dimension shape with a focus on minimising packing space gave rise to the double helix twisted spiral stair case like structure. Sequence of amino acid code information in a DNA structural form enabled easy duplication, which at the same time, allowed itself to be deciphered for assembly of amino acids.

The DNA became the information storage mechanism for propagation of itself and in cohort with RNA, built a better system to produce proteins from amino acids. This mechanism of protein manufacture using a system of data storage, replication of information and adopting a totally fixed universal code for the amino acid gave rise to a perpetually sustainable system that had no end game. The developed replicating system along with the protein synthesis mechanism became an unstoppable adventure just like a nuclear reaction that has gone haywire. There was no stopping of its path to glory.

CHAPTER 13

DNA-RNA—the
Only Successful Platform

The replicating mechanism and the system, i.e. the RNA, amino acids, protein making, DNA information data warehousing, duplication of RNA and DNA and the information flow between these molecules formed a perpetual cycle of forming and manufacturing each other serving the purposes for each other. It is a system of them, by them and for them. This self perpetuating cycle triggered diversification of life forms. Apparently, no other alternative molecular structure emerged in nature like RNA—DNA system on Earth. The early days must have eliminated those other wannabes. Ultimately, there was only one singular channel of trials that succeeded. The path of RNA and DNA was naturally selected, and RNA-DNA world became the only standard on Earth.

This DNA-RNA-Amino Acid-Protein mechanism became the uniform platform for all life forms on Earth. The platform did NOT change, the types of amino acids did not change, and amino acid coding nomenclature did not change. So, the entire molecular system for replicating and duplicating molecules ended up the same throughout the living world on Earth, with no exception whatsoever.

The coming together of amino acids, ribose, nitrogen bases and phosphoric acid to form precursor molecules to RNA, and DNA happened in a crucible within a vast expanse of watery ocean or maybe it could have happened in the hot water spring or in the semi colloidal clay along ocean fringes or in the depth of oceans. But it did happen somewhere on Earth very long time ago.

Our Earth started cooling down some 4.6 billion years ago from the whirling dust of exploding stars after the Big Bang some 13.7 billion years ago. The Earth was a molten torch of fire crust with an atmosphere so different from what we see today. The raging earth ball of fire exposed to the surrounding cosmic ocean from the fall out of the Big Bang with incessant bombardment of debris, meteors and fragments of star remnants, was undoubtedly a different sort of place to start off with.

The distance of Earth, at a strategic location, not far from our nearest star, the Sun, enabled water to exist as liquid when the early Earth cooled down. Earth's gravity and atmosphere aided in the formation of oceans and inland seas. The Earth was in a continual flux of formation from the molten rocks. The cooled molten solid crust floated on the molten lava. The fragmented crust moved on the liquid core as we see it today as movement of plate tectonic. The process of cooling is continuing still and is still in the process of flux, continuously adjusting to the movements of the plate tectonic.

During the flux, of the constantly changing condition of early Earth some 3.5 billion years ago, one successful crucible delivered amino acids, ribose, RNA and DNA. It took millions of years for that one crucible somewhere in the formative Earth. The possibility of one crucible getting it right with the right ingredients over 500 million years is not improbable. Sufficiently a long time period, I think, for the molecules to meet each other within the confines of this small Earth, till they found themselves as replicating molecules to further itself to unknown course.

The basic life system emerged at least 3.5 billion years old. This singular life template system became the normality for all living things on Earth. All life forms on Earth used this very old basic core system. The basic living system that emerged 3.5 billion years ago is the ancestral common framework which all living creatures, plants and bacteria including us share. We are all the products of same the molecular adventure of the elements.

The cell containing the vital molecules crafted itself as an independent, self sustaining mechanism. The cell membrane was a significant step forward in assisting the molecules to contain itself in a capsule form which allowed it to carry out functions within a desired environment which it could control against the vast wilderness and harsh environment in which it was thriving. They got it made, their own world separated from ocean full of other unwanted materials, substances and chemicals. This paved the way forward for a grand adventure. The cell invented the system of cell division instead of reinventing itself to propagate. The entire system was duplicating and not just the molecules.

The whole Earth became the playground for the cellular entity with aimless direction and with no ground rules or future course. Perfection was not its game, trial and error was. They existed for the sake of existence. It was elemental carefree play. AND time was of no consequence for this system to evolve.

With the cellular structure, the RNA-DNA-Amino Acid-Protein mechanism established itself as, the, and only platform for proliferation of life on Earth.

CHAPTER 14

The Masters of DNA

Early life forms used chemical energy to spur its activities, and it eventually looked at sunlight as a source of energy as it was available in abundance everywhere. But this energy needed to be captured and stored for reuse. Nature drove the cells to look at the available molecular resources such as carbon dioxide and water to produce energy molecules. The adventure seeking cells learnt the art of harnessing the energy from the sunlight through a process called photosynthesis and used water and carbon dioxide as raw material to produce carbohydrates as energy storage molecules.

Photosynthesis was an inevitable process devised by the cell. This process by default released Oxygen as a by product from the water molecule. The atmosphere during that period of development did not have significant Oxygen content and early life forms such as algae and bacteria enriched the Earth's atmosphere with Oxygen through photosynthesis while the cells were aggressively storing the produced carbohydrates.

The Photosynthesis Process

CARBON DIOXIDE (CO_2) + WATER – SUNLIGHT ENERGY $\rightarrow$ CARBOHYDRATE + OXYGEN (O_2)

With photosynthesis established, Oxygen began its foray into the atmosphere. Oxygen was not part of the environmental equation originally but it posed a new dawn on life forms. The cells went changing the environment of early Earth and in turn the environment that it changed posed a major challenge to its very existence. The new environment began to influence the course of the cell's adventure. Meanwhile, the cells were mindlessly producing carbohydrates and were fine tuning the apparatus for photosynthesis to produce even more energy stockpile.

As these cells proliferated, the Oxygen content in the Earth's atmosphere began to increase. The DNA had to endure and learn to live in this new oxygen rich environment. The harsh environment enslaved the DNA and pushed it to develop new methods to mitigate the developing environment in the atmosphere. Certain opportunistic cells and DNA started to look at the overwhelming situation of excess oxygen. The environment began to manipulate the DNA and cellular processes. Having built the structural components for photosynthesis, the DNA had the knowledge and skill to further indulge in making other mechanism based apparatus.

There was plenty of stored energy, and the opportunistic cells looked at using the chemical energy already captured and stored from photosynthesis. If the cell could use the stored energy, it will have the option to dispense with the photosynthesis mechanism if it wanted to.

First thing first, the cell had to learn how to harvest the energy stored in the chemical bonds of carbohydrates. The cell had knowledge of photosynthesis mechanism and this learnt mechanism was used. To unlock the energy, all the cell had to do, was to find the way to do the reverse of photosynthesis. Reacting carbohydrate with oxygen will yield energy. A simple reversal of the energy harvesting process!

Reversal of Photosynthesis—energy releasing process

CARBOHYDRATE + OXYGEN (O_2) ➔ CARBON DIOXIDE (CO_2) + WATER (H_2O) + ENERGY

Separates apparatus for the reverse energy releasing process was developed. The skilled DNA further learnt to craft itself to the ever changing demand of the environment which dictated the adventurous path of DNA diversification.

Before the proliferation of photosynthetic cellular forms, oxygen was not abundantly available in the early Earth environment and the living systems used what had been available in the surrounding, such as compounds of nitrogen, compounds of sulphur for their energy needs. With the oxygen enriched atmosphere, energy releasing mechanism using oxygen and carbohydrates became the overwhelming choice, and this gave further impetus for cellular diversification.

The power generating energy station was a crucial apparatus the cell crafted. The powerhouse biological equipment was modelled along the same line as the energy capturing apparatus for photosynthesis. The apparatus for capturing energy from sunlight and subsequently another device for releasing energy stored in chemicals were developed over 500 million years. I suppose the time was well spent in acquiring such facilities within the cellular portfolio.

Chloroplast was the apparatus crafted for capturing energy, and the apparatus constructed for releasing energy was mitochondrion. They existed as organelles in the cell. They would have existed as singular bodily entities on their own in the early Earth as they had their own set of DNA. The cell incorporated the organelles without having to reinvent the wheel so to speak. Available resources, biological equipment and fairly well working systems were selected and incorporated as the cell developed. It made use of all available resources.

The environment within which the DNA dwelled and the outside surroundings were dictators and masters of DNA. DNA did perform a good orchestra in concert with the environment and continues to do so. After all, DNA is a child of the natural environment.

CHAPTER 15

Cell Body & DNA Orchestra

The cell equipped with RNA, DNA, Chloroplast, Mitochondria and supported with biological processes made a giant leap forward to further adventure into the unknown. Biological instructions and mechanism were fine tuned over a period of more than 1 billion years. The cells were getting adept at it and were continuously adapting to the ever changing environment and did not fixate into a permanent mould. The DNA learnt to change and adapt to any change in the environment. The DNA had all the information data bases captured whenever there was a change in the environment and it subsequently incorporated the changes in the cellular structures via protein synthesis mechanism through the amino acid coding system.

As the cells got more organised, the apparatus in the cells began to compartmentalise, and the

DNA molecules were enveloped in the central core surrounded by the other materials of the cell. Chloroplast and mitochondrion were packaged as organelles in the cell together with their own specialised DNA molecules.

It took about 1.5 billion years for a simple primordial cell to organise itself into an independent cellular structure with a well defined nuclear core together with other organelles. It did take its time, a very long time developing from a simple cell to a more defined organised cell. It fine tuned many structures, processes, acquired apparatus and cellular, biological infrastructure, while adapting to the changing environment. It was not a planned activity but was a consequence of environmental requirements. Picking, selecting, eliminating, modifying, amending and changing the cellular body forms, changing apparatus and in concurrence changing the DNA databases and amending protein requirements, was its task. Biological processes were fine tuned for the cell to function as a complete self contained moiety that was able to duplicate itself and share the replicated copy of DNA. The ability to duplicate was passed on to the cell. Self replicating molecule organised itself into self duplicating cells. The information data base remained within the DNA, and the rest of the cell was only the body form that protected the DNA material and efficiently assisted in the proliferation of the DNA along the course of its adventure. The DNA moved on with its job of proliferation and did all that was necessary to propagate, by enlisting the support of cell's apparatus which, functionally supported the activities needed for its own replication.

Changes in the DNA molecules were reflected through changes in proteins. Any change in the DNA eventually manifested itself in the cellular behaviour, form and shape. These changes were passed on when the double strand DNA divided during cell division. The DNA just went on duplicating with the changes and added nitrogen bases for good or for worse. DNA was equipped to carry on its unplanned journey. The organised cell knew no boundaries and it went on with a wild determination to propagate itself. A few molecules that started 3.5 billion years ago organised themselves as cells after about 2 billion years. It developed various bio chemicals, reaction pathways, synthesising abilities, energy capturing protocol, energy molecules, energy releasing mechanism and logistics systems within the cellular form. Meanwhile, the information power house, DNA, kept developing the required molecules, chemicals, proteins as and when they were required.

Strangely and surprisingly, the DNA formula never changed, the type of amino acids 20 of them, the 5 nitrogen bases, the 3 system nitrogen base codes for amino acids, remained unchanged since. Even the chloroplast and mitochondria had no other alternative replacement system.

There was no other formula for duplicating molecules observed on Earth. There was simply no alternative to RNA or DNA or to amino acids or nitrogen bases.

And why was that?

Possibly, once the RNA and DNA formed in tandem with amino acids, the proliferation rate was overwhelmingly high, and the system singularly focussed on propagation rather than anything else giving no chance for any other alternative method to emerge. In any empirical case, getting a singular system itself organised was already a mega leap forward. The molecular march continuously chose the same system which became the established model. It was a good model hard to come by, anyway. The coordination between DNA and the rest of the cell was excellent as each component was feeding the other with new learning, knowledge and capacity building. The harvesting and use of energy mechanisms gave the Cell Body—DNA partnership the mobility to go anywhere it wanted as a combined entity.

What the cell needed was raw materials, chemicals, direct sunlight, access to easy availability of oxygen and space to move. When these were available in abundance, the whole world, so to speak, was its' oyster.

Fuelled by a continuous sunlight energy source, cell propagation was fast, and the cells began to form colonies amongst themselves, paving the way to cell specialisation and differentiation and coordination within the cell consortium which eventually led to multi-cellular organism in the ocean. Different environment gave rise to differentiation of the multi-cellular organism into various bodily shapes and sizes. Different environmental condition suited development of different body forms. But all these different type of body forms were based on the same platform following the dictates of DNA. The DNA modified itself as it went along in the differentiation of cellular type, giving new body form organism a better chance at survival in the ever changing environment. Simple division of body form was not viable anymore in multi cellular organisms. New method of propagation was developed. Specialised germ cell was the outcome together with separation of sexes, which allowed for exchange of DNA information database amongst it similar body forms. Separation of sexes and death was invented as a consequence. The germ cells took the path of self survival and propagation while the somatic cell (the bodily form) became the discard able used biomass.

With the changing demands of the environment over billions of years, the different body forms kept adapting to the new challenges filling the oceans and seas with numerous, diversified, specialised body forms that were able to propagate and change its forms in newer frontier when they moved into unfamiliar ground. The same cooperation and coordination between cell and DNA was reflected in the Cell-Body Form-DNA collaboration.

CHAPTER 16

Incidental DNA for *Homo sapiens*

C ell diversification, together with corresponding changes in the DNA, was prolific, and the changes were rampant. The oceans and seas were major playgrounds for developing various species. Propagation was paramount and the DNA molecule underwent massive changes frequently. The body forms were functional to the environment and many accessories were added on and incorporated. Gills, fins, shells, protective layer, colours, sensory systems and neuron systems for memory storage were developed alongside other essential appendages for survival. Food resources were not only limited to vegetables and plants, which were first level energy storage reservoir, but other body forms were also seen as secondary sources of energy molecules and amino acids. This introduced prey-predator concept.

The food chain developed while associated skills and warfare mechanism were nurtured for protection and hunting.

Vegetation and moving life forms developed prolifically in the seas and oceans while adapting to the varying environment. Specialisation of various cells for the harnessing of energy, logistics, and reproduction were enhanced further and the new frontier surrounding the water world was seen as a good option for the vegetation to move on in pursuit of better coverage of sunlight harvest. The same DNA platform was used when it ventured into land. The mechanism of protein synthesis, amino acid coding, apparatus for energy capturing, energy utilisation and corresponding processes did not change except for the modification of DNA and body forms which were dictated by demands of the environment, immaterial where it went.

Plant forms eventually colonised land and flora diversification went beyond control. The colonisation of land by plant vegetation, which were highly specialised for solar energy harvesting, was indeed a major triumph for the DNA lineage. Variety of plant forms emerged suiting to different land environment, from dry to wet terrains, hot to cold temperatures. Various adaptations were made, and the biodiversity grew.

The plant domination of land paved the way for moving life forms in the seas to step in as resources were plentiful. The body form needed modification and the DNA needed to acquire new information bases as the environment demanded its foray onto land.

The body life forms in the sea changed some of its apparatus and accessories to suit the requirement of the new environment on land less than 1 billion years ago. Other essential functions for survival such as nervous systems, respiratory systems and thinking mechanism were further enhanced to seize opportunities in this newly found land. The adaptation did take its time, but once on land, the moving body forms were going berserk as food resources were aplenty and their easy mobility on land was a crucial factor in their proliferation, diversification and reproduction. The moving life forms were conquering the land and were adapting their body forms to different environmental conditions over millions of years.

And along the path of diversification came the thinking animal. Homo like species appeared only in recent history some 5 million years ago. Common line of ancestors adapted to the changing environment by moving from various semi—human varieties to *Homo sapiens* (modern man). This development of Homo species was only a brief adventure of DNA as compared to the entire struggle of DNA over 3 billion years. *Homo sapiens* is only an incidental result in the adventure of the molecules that was lurching some 4 billion years ago in early Earth doing what was best for itself. *Homo sapiens* is not a planned organism in the DNA world, as DNA itself was an unplanned molecule. The DNA platform nurtured over 3 billion years is the same in Man and Man shares the same common ancestral birthplace with the rest of the fellow life forms on Earth. Man had many ancestors along the life tree right from primordial bacteria to primates and every other life form on Earth, some are close, and some are far relations but nevertheless share the ancestral DNA molecule, which is the platform for all life forms on Earth. Natural Selection and Environment were great shapers of history in modifying the DNA molecule. Without the continuous changes in the DNA, body forms and the changing environment, many organisms could not have emerged or survived. Many did emerge but failed to survive due to environment reasons and natural selections. Man's ancestors survived the harsh dictates of the environment and paved the way for humanoid to emerge. Due to the relentless march towards efficiency for a better equipped surviving machine, many predecessors of the Homo species developed the thinking mechanism and sharpened it along the way which finally culminated in *Homo sapiens*. Thanks to all the Fish, Amphibians, Reptiles, Mammals, Primates, Homo species and the rest who helped us along the way and also gave us a significant thinking apparatus and their related accessories. Man shares the same ancestral birthplace with other very humble early Earth fellows who developed the successful DNA molecule.

DNA is an earthling molecule and Man is but a recent new comer to the scene on Earth. Various body forms such as microbes, algae, plants, fishes, worms, mammals, marsupials, amphibians, reptiles, insects, birds, viruses and bacteria continue to modify and survive in a diverse environment over billions of years through natural selection and through evolution for the purpose of self propagation. The same principles continue to operate on all living things on Earth till today and will continue to do so. The adventurous nature of the DNA will not change.

Note: It is not the intent of this book to delve into evolution which is beyond the envisaged scope of this book. Excellent materials are available now on evidence of evolution for the interested readers.

CHAPTER 17

The Next Level

The emergence of Homo species and the early likes of the species can be traced to a period some 5 million years ago. History of *Homo sapiens* goes to about 200,000 years and that of modern, civilised man, to about 10,000 years. A recently developed species, I would say, as compared to other long existing species like crocodile and turtle which are at least 200 million years old. The dinosaur lived for 160 million years, some 65 million years ago, and almost all branches of that species are extinct. More organisms existed and went extinct than the number of organisms that exist today. The DNA diversity survived numerous extinction waves that came in the form of natural calamities, bombardments from meteors and from climate change. Against, these trials and tribulations, the DNA, and its associate molecules driven by a purpose of self propagation made a success story through the cell. Earth was the playground for these molecules. Organisms, humble though as they maybe, battled the harsh environment to propagate and evolved to colonise uncharted territory and spared no means to venture into difficult terrain and atmosphere. This amazing biological system finally delivered Man's thinking apparatus. The advance made by Man during the last 10,000 years is testimony of DNA's unplanned foray into the unknown. This chemical is no more solely responsible for future modification or evolution or advancement of Man as it relegated the functions of information gathering, storage and analysis of data to specialised information storage and processing organ, the Brain. With this thinking mechanism, Man was able to analyse, communicate and plan, which DNA itself never acquired. The thinking apparatus, the brain, well endowed in Man, is a result of our ancestral lineage which fought itself to our station today. We owe it to the primates and the other Homo species that existed before us, some failed, and some disappeared. Natural selection crafted a better thinking apparatus in Man. Our early ancestors and living things on Earth gave us all the necessary apparatus and biological equipments including the brain. We are an ensemble of coded knowledge of all the experiences, learning and trials of DNA from our various common ancestors including the primordial ones.

The DNA never had any pre plan, but its intention has always been the same, to duplicate, modify itself and propagate. It has no other objective. It captured the high sky, the water world and the rest of the Earth and sure, is yearning to colonise more. DNA cannot think, but Man can. As the progeny child of DNA, we must venture beyond the confines of our comfort zone, Earth. The journey of this molecule should not stop or halt here.

CHAPTER 18

High Cosmos

DNA is a one off molecule that emerged on Earth. The DNA molecule, with the amino acid coding system along with the accumulated databases for building cells, body forms, organelles for capturing energy, mechanism for release of energy, literally racked havoc on Earth.

A molecule assembled from basic available elements was able to determine the emergence of diverse life forms on Earth. Amazing, but given the length of time, that it had, the eventuality of what we notice now, though is a wonder, is not an inconceivable result. The elements were all there and available on Earth when it formed from the dust of exploding stars.

These elements are also abundantly common everywhere in the Universe. Then, the question remains whether there are any DNA type molecules in existence, in any corner, of this vast Universe?

The Universe as we see today is as old as the Big Bang itself, which happened some 13.7 billion years ago. The Universe is still expanding as the galaxies are spreading as a result of the Big Bang mega explosion.

Our home happens to be in the Milky Way galaxy amidst billions of galaxies in the Universe, and each galaxy has billions of stars which may host planetary systems just like our Sun. Our Sun is host to 8 planets and every star on a conservative measure can be assumed to have at least one planet. We are now discovering many more planets outside the Solar System.

On a rough estimate, the number of galaxies in the Universe can be said to be around 100 billion, i.e. 100 X 1000 million = 100 X 1, 000 X 1,000,000= 10^{11} galaxies. Again as a highly moderate estimate, the number of stars in each one of these galaxies can be assumed to be around 100 billion stars i.e. 100 X 1000 million stars. Therefore, the estimated total numbers of stars in our 100 billion—galaxy Universe would be 10^{11} X 10^{11} = 10^{22}. Difficult to fathom, but, someone said that there are as many stars as there are sand grains on Earth's shore. If one star hosts one planet, we may have that many numbers of planets, at the least! Whatever, it is a colossal number of stars and planets to deal with.

Our home galaxy, Milky Way, is in some remote corner of the Universe. The Milky Way Galaxy is around 100,000 light years across which means, it will take 100,000 years to travel from one outlying area of the galaxy to the opposite point, that is, if we travel at the speed of light. Milky Way is around 2.5 million light years from our neighbouring Andromeda galaxy, and the observable Universe from Earth is said to be around 92 billion light years across. It will take 92 X 1000 X 1,000,000 (9.2 X 10^{10}) years to cross the Universe if we travel at the speed of light! One light year is 9,460,730,472,580 km (~6 trillion miles) in distance, and we have to travel 9.2 X 10^{10} X 9,460,730,472,580 km to get the glimpse of our Universe. This is an unimaginable space and distance to comprehend. It is mind boggling for our earth grown brains.

Whether you like it or not, we are not able to travel at the speed of light with our current technology. Even surveying our local Solar System within the Milky Way Galaxy is already an immensely challenging effort.

We have landed Man on the Moon and our next target obviously is Mars. Mars is a close neighbour, and it is the fourth planet from the Sun while Earth is the third from the Sun. Given the current technology it will take about 6 to 10 months for a space craft to reach Mars. The distance at the nearest point between Earth and Mars is about 55 million kilometres.

Man, after having developed the biological systems which were assimilated from various common ancestors added on the thinking portfolio to the brain function, purely as an environmental necessity for survival. Man's advancement in knowledge acquisition outside the physical form and permanent recording ability made him an extraordinarily successful organism. Man, armed with the thinking mechanism and technology, is able to look beyond the shores of Earth and venture into the Cosmic Ocean.

We have sent unmanned mission to Venus, Mars and to Titan, one of the satellites of Saturn. Spacecrafts have flown by Jupiter, Saturn, Neptune and Uranus. A small exploration, given the vastness of space! We have even sent space robotic probe beyond Saturn and Jupiter, which went on a tangent off the Solar System with information about Earth and Man, DNA structure, music and what not. We hope some intelligent life form might pick it up one day.

We embarked on space exploration not only to understand the Universe and also to gain knowledge on environmental conditions of planets and moons out there. One day we hope to colonise. Man's adventurous ambition will remain uncompleted till we conquer the Universe.

However, space travel technology has a long way to go and Man's biological preparedness for a long period of space travel is questionable. Man, as an earthling is not biologically suited to travel in space, anyway.

In any case, are we alone in this Universe? Is there any life form elsewhere in some corner of the Universe? Do they share the same or similar chemical platform? Whatever, it will be an exciting new world if we find life forms in some part of the Universe. The landing of Curiosity rover on Mars will further push our knowledge frontier on the possible existence of exobiological molecules.

But there is this one obvious burning basic question—Is there any intelligent life form anywhere in this vast Universe? We wish there are!

But, how do we find them?

Advanced communication ability, is, but one key criterion, to qualify any life form as an intelligent being. Whether we can communicate with them, is a different issue all together. But if we are able to communicate, we will have a host of fundamental questions to ask, including whether they have the technology to travel at speed of light and whether they are aware of any other intelligent life forms elsewhere in the Universe? Learning the chemistry of their living molecules would be a giant leap in understanding life itself from a totally different perspective.

We will eventually find them whatever it takes if they exist!

Space adventure is a passionate pursuit of Mankind, and it will remain so. We should continuously seek signs of intelligent beings in the Universe while relentlessly looking for any life form primitive or otherwise which may be in different stages of development.

However, is there any other way of adventure into Space? Is there a better way of doing things?

Or can we do something else—very different?

CHAPTER 19

Expanding the Playground

Technology remains the greatest challenge to our desire to seek the unknown Universe out there. Travel through space-time warp or worm holes are concepts thrown around to facilitate travelling light years in a short span of time. Maybe, one day we will realise the dream, but now it is still in the realm of science fiction. But the quest to get out there to seek is unquestionably, the desire of all humankind. As it is dictated in our DNA, we need to go beyond the confines of our cocooned environment on Earth. The quest is perfectly in line with human endeavour and Man has this innate need to find out what else is there in the vast, unimaginable Universe for him to conquer and propagate.

Yes, we want to travel as far as possible from Earth to find out what's out there.

Obviously, we can make short trips to our local planets and moons in the Solar System. The environment may not be quite amicable. We need to handle all obstacles through technology. The challenges are humungous. We have sent unmanned probes beyond Solar Systems to capture data and transmit to Earth for us to analyse and plan next course of action. Basically, space travel has been an information gathering exercise. Though it increases our knowledge base, it does not give us a quantum leap in space exploration. Curiosity in Mars is obviously a significant leap forward.

Yes, we want to know whether there are any life form whatsoever in any of the planets near or far, near would mean a couple of light years and far would mean 100 over millions of light years away.

We can only examine pictures of terrains sent by space probes, but we will not be able to discern whether the stuff we see is living or non living without physical investigation. We may be able to detect signs of civilisation if the planet is inhabited by intelligent life forms. However, we need to develop criteria and guidelines for establishing exobiological existence, both for intelligent and basic life forms. What test method or criteria can we use?

We have only one example on the origin of life, that which was displayed on Earth. We have a thorough understanding of all the chemical processes and have identified the organic substances of earthly biological origin. We know of, only one Biology! Living chemicals elsewhere in the Universe may be equivalent molecules to that we find in biological system on Earth, but it may not be similar and for the matter may be totally different. However, the behaviour of elements throughout the Universe is absolute. The electrons follow the same rules everywhere in the Universe. Atoms are capable of forming living molecules anywhere in the Universe given the right conditions. So, occurrence of life forms in the Universe is a definitive probability.

Life forms whether existing or extinct would have used the same raw materials of star dust to build their exobiological systems. Those raw materials would be the same elements as in the periodic table. But the molecules of life or biological molecules cannot be expected to be the same or similar, even processes would be altogether different, perhaps beyond our comprehension. We cannot expect even a remote resemblance to RNA or DNA. Do we look for carbon based compounds or should we look for replicating and duplicating molecules? Maybe we should look for living systems rather than earthy organic molecules!

Life forms need not necessarily be carbon based. Carbon did a fantastic job on Earth though. We cannot expect evolution and biological development similar to what we have seen on Earth. We cannot expect Earth like planets with similar atmosphere yielding carbon based propagation to have evolved in the same way as the organisms did on Earth. We may be shocked to detect life forms with quite different cellular and body structures with different biological apparatus and sensing mechanisms.

Life forms may be based on silicon or tin or lead or aluminium which may be sustained in a medium other than water, may be ammonia or liquid hydroxide sulphide or liquid methane or anything else. Existence of water need not necessarily indicate the existence of life forms. Water is merely one of the components in the arrays of molecules that worked together on Earth. It would be naive to assume that the same molecular adventure would or would have taken place in other planets. Water alone is not the prime mover for life to begin. We must get away from this concept that water is an absolute indication of life-A Earth based presumption.

Duplication and a system that enables replication and propagation could be a basic criterion to use to ascertain any material as a life form candidate. No pre conceived notion on chemical compositions, chemical pathways, mechanism for energy capture, etc. We may be in for a bizarre, shocking experience in discovering life forms elsewhere in the Universe. Or we may just miss it all together.

Technology apart, it will be painstaking going from one planetary system to another looking for life forms, literally, searching for the proverbial needle in the haystack, which may not, be there within the universal region, we are looking at. It may take many centuries to survey the local region given the prevailing technology we have on space travel.

The search, however, must continue while we look for better space travel technology including bending space time continuum, if it is possible.

But what if the life forms can communicate, then things can be a different ball game.

Yes, we want to know whether there is any intelligent life form capable of communicating, existing or have existed or likely to occur in the future in any one of the planets in any corner of the Universe.

Communication was a necessary evolutionary development in life forms on Earth—plants and animals communicate using different tools amongst their species and also across species either to trap or frighten the other. However, Man also built a sophisticated verbal sound system to communicate. Verbal languages, which developed into written forms, became powerful tools for Man's technological progress.

We need to look at common communication mode or system if we want to communicate with intelligent life forms, presumably also able to communicate. Basic communication is exchange of information or data. The transmitter and the receiver need to be aligned in the same wavelength so to speak. The data must make some sense to the recipient, as well. One is the content of communication i.e. data or information, and the other is the mean of communication i.e. how this information is sent and accordingly received.

Again, we are looking at distances which are light years away in order to get our information across. How do we transmit? What mechanism is available to send information fast in the Universe so that time is shortened for communication? The only known fastest speed in the Universe is the speed of light. The option we have is electromagnetic waves, which travel at the speed of light. This mode of communication is well established. The radio wave, which is part of the electromagnetic spectrum apparently, is the best option we have.

However, the critical question remains, whether the other intelligent life forms receiving the data have the capability and technology to capture, decode and interpret the data or information-should the information be in symbols or graphical pictures? Can they see or hear?

The intelligent life form should have the ability to comprehend information received at that material point in time when they receive the data. Otherwise, they would miss it, if the data reaches them when they have no technology or when their civilization has vanished. But it would not mean that there are no intelligent life forms in the Universe if we do not receive any response whatsoever!

If there are intelligent life forms elsewhere in the Universe seeking out other life forms, targeted their transmission to our part of the Universe, then the transmitted data must be receivable and readable by us. If we are unable to receive their kind of signal which would be different, then, we would miss the opportunity as well. Timing is vital as the Universe is vast and the transmitter and receiver life form communities should be co existing with mutually competent technology during the transmission. Though the probability of existence of life forms is high, chances of two life form intelligent civilizations coming together with the mutually compatible technology at one material time may be slim.

Nothing would have happened if some intelligent life form, say about 100 light years away from Earth sent information through radio wave some 1 million years ago hoping for a response. Obviously, there would have been no response from Earth and the sender after 200 years would have assumed that there are no intelligent life forms in our Solar System region and would have aborted all attempts. In any case, in this example, it would take 200 years for an exchange of information between Earth and some planet 100 light years away. 100 light years is a common local area in the Universe. What more if the intelligent life form with the capacity to communicate is a little bit further away, like 1000 light years away?

The fastest means of communication we can achieve under current laws of physics is the speed of light. Even if we are fortunate to find one intelligent life form in some corner of the Universe, the time lapse for two way communication would be too long. But we shall nevertheless, not give up and shall endeavour till the end of our civilisation.

The probability of existence of life forms or intelligent life forms is high, based simply on the sheer number of stars which are capable of hosting planetary systems. However, the prospects of meeting life forms or communicating with them appear sadly remote.

But if some of the advance intelligent life forms have found the way to travel, then there is a chance of these beings visiting us if they found our part of the region, exciting to them. Or we could have been deemed as a bunch of irrelevant, unimportant life form existence, as there could have been plenty of more fascinating life forms found across the Universe in their vicinity. Or maybe we are not the priority life forms for their study. Maybe they have no allocation, of time and resources, to get in touch with us as yet.

However, in any case we should continue our search for extraterrestrial intelligence. But what information do we send to the intelligent life forms sitting some 100 light years away from us?

The aliens may not be able to see as we do, so pictures may not be recognisable. They may not be able to hear within the sound wave frequency and sending music of Mozart, Erhu and Carnatics music pieces or speeches in multiple languages may not help.

Well and good if they can see or hear or both. Possible but may not be probable.

So, what do we do?

Maybe it would be advantageous to send information in different wavelength within the electromagnetic spectrum. The intelligent life form may be able to pick up the information and data in the wavelength it is familiar with and be able to interpret.

Of all the information, we may want to send, the periodic table should take precedence. Conceptually it is universal as it occurs throughout the Universe and all life forms have to use the elements in the periodic table. Letters such as H or C or Na will not be comprehensible but symbolic representation of atomic number and electron configuration should be cleverly designed and transmitted. All advanced intelligent life forms would have mastered the workings of elements. If they can decipher the periodic table, we have won the battle in communicating. The intelligent life forms with established technology will have knowledge of the chemical elements and the electron arrangements. This is basic absolute chemistry. Without this knowledge, no intelligence life form can advance their technology. This is the common knowledge that will bind and bring bonding between intelligent life forms existing in the Universe. Everything else would be totally alien to each other.

The life form that could decipher the periodic table information and the logical arrangement of naturally occurring elements will come to the conclusion that the party that sent the information is indeed intelligent, as well.

Milky Way's coordinates in relation to other galaxies and the Solar System are indispensable directional information for alien life forms. However, data on DNA, amino acids, languages, biological life forms on Earth are secondary, a good to know information, which can be exchanged at the later stage. But initial understanding is vital. And, any simple response from any other life form would be a galactic leap forward and would be the pinnacle of Man's celebration of his own existence.

CHAPTER 20

The Adventure is Unstoppable

T he elements and molecules had a field day over the last 3.5 billion years going through relentless trials and tribulations in its march towards putting together a replicating molecule and developing an entire system of living flora and fauna on Earth through natural selection in the midst of a violently changing Earth terrain and atmosphere. The raging Earth, moving tectonic plates, violent eruptions of lava, earth quakes, dramatic climate changes, atmospheric changes, were a continuing phenomenal challenge that our chemical elements faced and moved on-A success story against immeasurable odds. Many species disappeared and went extinct over the cause of violent, external changes and natural selection. The inadvertent and incidental emergence of *Homo sapiens* along the path of evolution is but a favour Man should eternally be grateful, because, without this evolution and natural selection, Man could not have emerged at all or perhaps could have but may have been a little less intelligent. The less intelligent species would not be wondering why it existed, where it came from or where it is going to or even wondered whether there are any other similar life forms like itself.

Undoubtedly, dolphins, whales, sharks, crocodiles, orang-utans, tigers, lions, elephants, eagles, snakes mudskippers and durian trees do not question its own existence.

Man's existence on Earth is only recent history. *Homo sapiens'* history in the whole realm of the Universe is rather brief, mere 200,000 years as compared to 13.7 billion years of existence of the known Universe.

The Universe will go on even after Man's extinction. Earth is not free from disasters and is exposed to external source of destruction through meteors and other cosmic events. In any case, when our Sun comes to the end of its life cycle, it will either engulf our Earth in the process or send the Earth on a tangent to nowhere. In the natural cosmic world, the seemingly peaceful Earth is not a safe haven.

We need to find alternative homes and colonies either in the nearby neighbourhood or far out in the outskirts and beyond the known cosmic horizon, however, remote it may be. The journey, which started from the star dust, has to continue.

The adventure of the elements and their prized produce on Earth, the DNA molecule, must survive through any incident of calamity or destruction, for it will be lost forever and may not surface anywhere else in the Universe. The grand trials of DNA should not stop, and Man as a marvellous manifestation of DNA, is in a position, to propagate the ancestry of DNA beyond Earth, to further its adventure.

Space travel will eventually take Man to the new worlds of planets and moons in distance stars. Humanity will marvel at the immenseness and beauty of the Universe and better understand the magnificent laws of nature. It may be a long wait that may never come.

Man, being a product of Earth, is ill suited to live in the unfamiliar environment of other planets. Intelligent as Man is, unimaginable risk mitigation plans would have to be put in place before the community ventures into sustainable space travel. But the progress of colonisation would be extremely slow, and there is no guarantee of success and Man is likely to face untold obstacles in unknown worlds and every step will be limiting on its own. But there is no choice Man must continue this path, the path less travelled. The proverb-it is better to travel with hope than to arrive, is certainly very apt in this scenario.

Man wanted immortality, but that is not possible as DNA's path to diversity transferred mortality from somatic cell (body cells) to the gamete cell long before Man's advent on Earth. The invention of sex and death was well crafted much earlier before the arrival of Man. Conceptually an individual man do not die, he merely passes on the gamete cell to the next generation, alas minus, the individual memory and physical body. The individual man lives on through passing of genes and the specific DNA molecules to the progenies. The entire human species lives and passes on the DNA and propagates except it takes on similar body form dictated and instructed by DNA.

So it actually does not matter whether an individual man travels or his genes or the DNA goes on this great sojourn of the Universe to colonise. In fact, DNA is better suited and more flexible in handling any harsh environment than physically bodied man. DNA already has the experience, and with all the accumulated learning, it will be able to adapt itself to the new world in due course.

It would be a fascinating preposition to send DNA from selected species to targeted planets, as experimental stations, within and outside the Solar System. This is not contamination of planets but colonisation of earthly DNA. There is enough number of planets in the Universe and selecting a handful of suitable planets for this experiment would not adversely affect the nature of the Universe as such.

Perhaps, one day after millions of years, those DNA molecules would have adapted to the new planets. If life form appears in this experiment, then it will truly be a giant leap forward, what more, this emerged new life form will have the same ancestral roots to earthlings, though we can't expect much similarity. And, if it had evolved into some intelligent life forms, we will at least have a chance to communicate with them.

The continuing search for intelligent life forms through SETI (Search for Extraterrestrial Intelligence), space probes, manned and unmanned space exploration are endeavours in the right direction, but it may take millions of years before Man be able to witness intelligent life form other than *Homo sapiens* or even come across a basic life form, not because it is not out there but simply because we just do not have the capacity to do so against the challenging vastness of the Universe.

I wonder what our earthy DNA is capable of beyond its own domain! Given some assistance along the way, the DNA molecule, I believe, will reinvent itself rather swiftly without having to struggle through the initial trials. It will just have to reconstruct itself in a totally alien terrain. It would be a quicker learning curve for the DNA. But what it will become is anybody's guess, as its journey has always been unplanned. Whatever, it is a very capable molecule. After all, it did produce you and me.

The chemical elements and its adventures are beyond our imagination, and its manifestation on Earth though is grand, is just one mere possibility. While Man was never the prime mover in life itself, he can, however, dictate how life can become. Life is not about Man's journey, it is about DNA's journey and as an intelligent life form we owe it to the DNA molecule, to adventure and propagate.

The elements are with us for this adventure, and if we choose this path, the electrons will take us to places we have never been before.

We may have the END, but there will be no END for the adventurous elements and molecules in this Universe.

ABOUT THE AUTHOR

Venthan Nalathamby was born in Pasir Penambang, Malaysia in 1957. He graduated with a Bachelor of Science (Hons) degree in Chemistry from University of Malaya, Kuala Lumpur. He has a passion for Biochemistry and hopes to pursue his interest in Astrochemistry.

www.ingramcontent.com/pod-product-compliance
Lightning Source LLC
Chambersburg PA
CBHW032015170526
45157CB00002B/711